ENVIRONMENTAL EPIDEMIOLOGY
Exposure and Disease

Edited by

Roberto Bertollini, M.D., M.P.H.
World Health Organization
European Centre for Environment and Health
Rome, Italy

Michael D. Lebowitz, Ph.D., FCCP, FACE
Professor of Medicine (Pulmonary) and Epidemiology
University of Arizona
Tucson, Arizona

Rodolfo Saracci, M.D.
International Agency for Research on Cancer (IARC)
Lyons, France

David A. Savitz, Ph.D.
Department of Epidemiology
School of Public Health
University of North Carolina
Chapel Hill, North Carolina

Published on behalf of the World Health Organization Regional Office for Europe by Lewis Publishers

LEWIS PUBLISHERS
Boca Raton   New York   London   Tokyo

Library of Congress Cataloging-in-Publication Data

International Workshop on Priorities in Environmental Epidemiology
(1993 : Rome, Italy)
Environmental epidemiology : exposure and disease : proceedings of
an International Workshop on Priorities in Environmental
Epidemiology / organised by the WHO European Centre for Environment
and Health, Rome Division ; edited by Roberto Bertollini ... [et
al.].
p. cm.
Includes bibliographical references and index.
ISBN 1-56670-067-1 (alk. paper)
1. Environmental health--Research--Congresses. 2. Environmentally
induced diseases--Epidemiology--Research--Congresses.
I. Bertollini, Roberto. II. WHO European Centre for Environment and
Health. Rome Division. III. Title.
[DNLM: 1. Environmental Exposure--congresses. 2. Environmental
Health--congresses. 3. Epidemiologic Methods--congresses. QZ 57
I61e 1995]
RA566.26.I58 1993
615.9'02--dc20
DNLM/DLC
for Library of Congress   95-17470
                                                                                                      CIP

The views expressed in this publication are those of the author(s) and do not necessarily represent the decisions or the stated policy of the World Health Organization.

The World Health Organization Regional Office for Europe welcomes requests for permission to reproduce or translate its publications, in part or in full. Applications and enquiries should be addressed to the Publications unit, WHO Regional Office for Europe, Scherfigsvej 8, DK-2100 Copenhagen Ø, Denmark.

© World Health Organization 1996
Lewis Publishers is an imprint of CRC Press

No claim to original U.S. Government works
International Standard Book Number 1-56670-067-1
Library of Congress Card Number 95-17470
Printed in the United States of America 1 2 3 4 5 6 7 8 9 0
Printed on acid-free paper

Foreword

The development of environmental epidemiology has been understandably slow: studies often involve very small changes in incidence of health effects compared to the variability that may be expected for other reasons. Neverthe-less, an increasing number of successful investigations have been conducted during recent years. Such work can have immense value in achieving a better understanding of the relationships between environmental conditions and health effects and in improving the basis for rational decision-making.

Environmental health is a very wide field, encompassing all those aspects of human health and disease that are determined by factors in the environment. The term, as used by the WHO Regional Office for Europe, includes both the direct pathological effects of chemicals, radiation, and some biological agents, as well as the effects (often indirect) on health and well-being of the broad physical, psychological, social, and aesthetic environment, which includes housing, urban development, land use, and transport.

Taking into account limitations in financial and human resources, it is essential that clear priorities are laid down for the most important subject areas for studies. The Workshop held in Rome in January 1993 provided an excellent opportunity for discussions among a group of experts on prioritisation within several aspects of the broad environmental health field. The conclusions and recommendations of the meeting must, of course, be read in context of *Concern for Europe's Tomorrow*, a comprehensive review by the WHO European Centre for Environment and Health of the whole range of environmental impacts of human health covering the 51 countries of the European Region. The report and the deliberations of the second Ministerial Conference on Environment and Health in Helsinki in June 1994 highlighted the significance within the European Region of, for example, microbiological hazards associated with contaminated water and food, as well as environmental factors linked to unsatisfactory housing and living conditions. It is clearly important that all aspects of environmental health should be taken into account when determin-ing which are the most important topics to be dealt with. The Rome meeting represented a very valuable contribution to this goal.

The Helsinki Declaration and its accompanying Action Plan have provided a major challenge to the Member States of the European Region. But rational and cost effective decision-making must also be underpinned by improved scientific knowledge. The organisers of the Rome meeting must therefore be congratulated on their initiative, which represents a most useful contribution to the continuing debate on how resources should be allocated and based, *inter alia*, on careful and objective risk assessment.

Sir Donald Acheson
London School of Hygiene and Tropical Medicine
London, U.K.

The Editors

ROBERTO BERTOLLINI

Roberto Bertollini, M.D., M.P.H., has been the Director of the Rome Division of the European Centre for Environment and Health of the WHO Regional Office for Europe since 1993. Before joining the WHO, he worked at the Epidemiology Unit of the Lazio Region of Italy.

Dr. Bertollini received his Medical Degree with Honours from the Catholic University of Rome in 1978 and a Postgraduate Degree in Pediatrics from the University of Rome "La Sapienza" in 1982. He then obtained a Master of Public Health from the Johns Hopkins University in 1983 and spent one year as postdoctoral fellow in Epidemiology at the same University in 1986.

Dr. Bertollini is a member of the Society for Epidemiologic Research, the International Society for Environmental Epidemiology, the Italian Epidemiologic Association and the Italian Pediatric Epidemiology Working Group. He is an associate editor of the *American Journal of Epidemiology*.

He has over 70 research publications in English and Italian and has given many presentations at national and international research meetings. Dr. Bertollini's research interests concern the environmental influences on reproductive health, pediatric health, and cancer, as well as the use of epidemiology for public health policy.

MICHAEL D. LEBOWITZ

Michael D. Lebowitz, Ph.D., is currently Professor of Medicine (Pulmonology and Environmental Medicine) and Epidemiology, Chair of the Epidemiology Graduate Interdisciplinary Program, and Associate Director (and Head of Environmental Programs) of the Respiratory Sciences Center in the College of Medicine at the University of Arizona, Tucson.

Dr. Lebowitz received his Ph.D. in Epidemiology and in Environmental Health Sciences from the University of Washington (Seattle). He also has a Ph.D. in Preventive Medicine from there, and a B.A. and M.A. from the University of California (Berkeley). He was also a National Institutes of Health Fogarty Sr. International Fellow in Clinical Pulmonology, Immunology, and Occupational Medicine, British MHLI/U, London, 1978–1979.

Dr. Lebowitz is a Fellow of the American College of Epidemiology, of the American College of Chest Physicians, and of the International Academy of Indoor Air Sciences. He is an honorary member of the Hungarian Health Society. He is also a member of the American Epidemiological Society, the International Epidemiological Association, the Society for Epidemiological

Research, the American Thoracic Society, the European Respiratory Society, the International Society of Environmental Epidemiology (ISEE) and the International Society of Exposure Analysis (ISEA). He has served on the ISEE and ISEA Executive Committee and various committees of other professional organizations.

Dr. Lebowitz has been a consultant to various U.S. and International agencies, including EPA, NIH, CEH/CDC, PAHO, WHO, and the EC/EU, as well as a consultant to various governments. He has been involved in various committees of NRC/NAS/IOM and WHO/EURO. He has had grants, cooperative agreements, and contracts from NIH, EPA, and other agencies and organizations. He has received several awards and citations.

Dr. Lebowitz was senior editor and contributor to the WHO monograph "Guidelines on Studies in Environmental Epidemiology" (EHC 27, 1983) co-author of the ATS "Health Effects of Air Pollution" (1978), co-chair and co-editor of the NRC/NAS "Indoor Pollutants" (1981), contributor to EPA Criteria Documents on PM/SOx and Ox, chair and co-editor of the WHO/EURO "Biological Contaminants" (1989), a member and contributor of the NRS/IOM/NAS Committee "Indoor Allergens" (1993); contributor to the WHO/EURO "Air Quality Guidelines for Europe" (1987) and author of a monograph for EPA and PAHO on "Methodologies for Health Appraisal Studies". He has co-edited journal supplements from various workshops and conferences and has been associate editor and member of several editorial boards, including the *Journal of Exposure Analysis and Environmental Epidemiology, Journal of Toxicology and Industrial Health* and *Archives of Environmental Health*. He has also co-published over 250 peer-reviewed articles and over 135 abstracts/editorials, has been contributing author of over 100 published official reports, and has conducted over 200 national and international presentations. His research has focused on air pollution, occupational exposures, and lung diseases.

RODOLFO SARACCI

Rodolfo Saracci obtained his M.D. degree with honours from the University of Pavia in 1960 with a dissertation on human genetics. He also holds postgraduate degrees in internal medicine (University of Pisa, 1965) and in medical statistics and biometry (University of Milan, 1971).

After clinical and research work in internal medicine at the University of Pisa, he moved to epidemiology as a visiting research fellow in 1964–1965 at the MRC Statistical Research Unit in London under the direction of Sir Richard Doll. He subsequently held posts at the Universities of Milan and Pisa, where he taught biostatistics from 1969–1976 and since 1976 at the International Agency for Research on Cancer (IARC) in Lyons, France, where he has been until the end of 1994 Chief of the Unit of Analytical Epidemiology. He is currently Director of Research for IARC, Lyons.

He contributed some 150 publications in English to the literature on the identification of carcinogenic environmental hazards, interaction between environmental agents, as well as on methodological aspects in epidemiology clinical trials.

He was a member of the Biometric Society review committee of the UDGP diabetes trial in 1972–1974, and he has been President of the Association des Epidemiologistes de Langue Francaise (1982–1985). He is President-elect of the International Epidemiological Association, Director of the courses of the European Educational Programme in Epidemiology (EEPE) and on the editorial boards of *The American Journal of Epidemiology, Annals of Epidemiology, Epidemiology, Cancer Epidemiology, Biomarkers and Prevention, Journal of Epidemiology* (Japan), and *Diabetologia*.

DAVID SAVITZ

David A. Savitz, Ph.D., is Professor of Epidemiology at the University of North Carolina School of Public Health, where he also serves as the co-director of the Occupational and Environmental Health Program, a Fellow of the Carolina Population Center, and is affiliated with the Lineberger Comprehensive Cancer Center.

Dr. Savitz received a Masters degree in Preventive Medicine from Ohio State University in 1978 and a Ph.D. degree in Epidemiology from the University of Pittsburgh in 1982. He joined the faculty of the Department of Preventive Medicine and Biometrics, University of Colorado School of Medicine in 1981 where he served as Assistant Professor until moving to the University of North Carolina in 1985.

Dr. Savitz is a member of the Society for Epidemiologic Research, for which he previously served as Secretary Treasurer and is currently a member of the Executive Committee. He also is a member of the American Public Health Association, International Society for Environmental Epidemiology, Society for Pediatric Epidemiologic Research and Bioelectromagnetics Society. He is co-chair of a U.S. National Academy of Sciences Committee addressing possible health effects of electromagnetic fields and a member of a National Academy of Sciences Committee on Persian Gulf War health effects. He is an Editor of the *American Journal of Epidemiology* and an Associate Editor of *Environmental Health Perspectives*.

He has over 100 research publications and has made numerous presentations at national and international research meetings. Dr. Savitz is engaged in a number of research activities concerned with environmental influences on reproductive health and cancer, as well as studies of nonenvironmental influences on reproductive health outcomes.

Introductory Remarks by the Editors

Environmental epidemiology is a rapidly growing field. The assessment of risk to human health from exposure to environmental hazards is often based only on toxicological evaluations; epidemiological data to support these observations are often lacking due to their complexity and cost. The best use should be made of resources, and research should concentrate on issues identified as relevant to public health.

The WHO European Centre for Environment and Health (ECEH) recognised the need for open discussion by experts in environmental epidemiology and related fields to identify areas of priority research in environmental epidemiology. An International Workshop, *Setting Priorities in Environmental Epidemiology*, was therefore organised by the Rome Division of the ECEH, with the help of the Lazio Region Epidemiology Unit (OER) and the Higher Institute of Health (ISS) of Italy. The workshop was held in Rome from January 28–30, 1993 and was sponsored by the OER.

The objective of the workshop was to identify priorities for research relevant to public health in selected areas of environmental epidemiology. The initial step was to review the available epidemiological evidence and agree on criteria to use in setting priorities. This book collects the working papers provided by the participants of the workshop. The workshop could not be comprehensive, but was aimed at establishing a consensus among experts on selected issues. The working papers covered the various areas of environmental epidemiology from three different points of view: environmental exposures, epidemiological methodology, and major disease groups related to the environment.

CRITERIA FOR SETTING PRIORITIES

In identifying priorities for research in environmental epidemiology, a set of criteria were used relating to exposure conditions, health effects, and the feasibility of human studies (Table 1). The first two areas received more emphasis; study feasibility was only considered when high priority was given to exposure or health effects. Thus, priority setting was based on both public health relevance and opportunities for research.

With regard to exposure, some items in Table 1 need clarification. Meeting the specifications implies high priority. An exposure affecting large numbers of people will in general result in a higher priority than one involving only a few people. Small groups, however, such as the people in certain occupations, may sometimes receive high exposures that result in substantial individual risks, and this also needs consideration. An increasing trend in exposure will

Table 1 Criteria for Setting Priorities in Environmental Epidemiology Research

Criterion	Specification
Exposure	
Frequency of occurrence	Large population exposed
	High percentage of population exposed
	Wide distribution of exposure
Level of exposure	Biologically relevant level of exposure
Trends in exposure	Increasing occurrence of exposure
	Increasing intensity of exposure
Health effects	
Harmful effects	Experimental data indicate harmful effects
	Biological plausibility
Available evidence in humans	Epidemiological evaluation not conclusive
	Exposure–response relation data are insufficient
Individual susceptibility	Effects on sensitive subgroups can be evaluated
	Potential synergism with other factors
Severity of problem	High mortality, morbidity, disability, cost

often lead to higher priority, but a decrease in exposure in connection with interventions may create a useful opportunity for environmental epidemiology studies. Other important circumstances include the duration of the exposure and the number of sources of exposure.

The criteria are based on public health issues, that is, the incidence, duration, and severity of health effects. The identification of sensitive subgroups may be important in understanding etiological mechanisms and maximising preventative measures. The participants realised, however, that the identification of susceptible individuals may also have ethical implications for genetic counselling and selective prevention, for example. Public concern may also influence priority setting to some extent. The priority given was lower than possibly merited if sufficient work was underway or completed.

Some criteria for study feasibility must be met, such as having an adequate time since exposure for effects to be observed prior to conducting a study. Other important criteria are the possibility of getting precise and unbiased data on exposure and disease (or other outcomes) and on confounding factors. Regardless of the funds available, any study should be carried out as efficiently as possible.

PRIORITIES FOR RESEARCH IN ENVIRONMENTAL EPIDEMIOLOGY

During the workshop, three working groups were established to discuss and agree on priorities for the following exposure categories:

- air contaminants, including airborne contamination from hazardous waste sites, asbestos, benzene and other carcinogens, CO, lead, NO_x, SO_2, and total suspended particulates;

- water contamination, including waterborne exposure to substances from hazardous waste sites (arsenic, asbestos, chemicals of agricultural, commercial, domestic or industrial origin, disinfection byproducts, fluoride, nitrates, and radio nuclides), and pesticides;
- ionizing radiation, electromagnetic fields, and exposure subsequent to manmade disasters.

The priorities identified by the three working groups are summarised in Table 2. A detailed description of the considerations which guided the meeting's participants in the selection of the priorities has been published elsewhere (*Arch. Environ. Hlth.*, 49 (4), 1994, p. 239–245).

Table 2 Priorities for Research in Environmental Epidemiology by Exposure and Disease Group

Exposure	Disease Group	Issues To Study	Level of Priority
Air contaminants[a]	All of interest	Exposure assessment methods.	High
		Siting of samplers for routine air pollution monitoring.	High
		Methods for respiratory and cardiovascular disease surveillance and registration.	High
		Use of diaries for acute or short term effects.	Medium
		Use of biomarkers for mix of exposures.	Medium
	Respiratory system (other than cancer)	Incidence and development of various stages of asthma.	High
		Incidence of diseases of the lower respiratory system in children, especially before age 6.	High
		Development and exacerbation of chronic respiratory disease.	High
		Acute respiratory diseases in people over age 6	Medium
		Lung growth and decline; development of bronchial reactivity	Medium
	Other diseases	Cardiovascular disease (in relationship with CO and ETS)[b]	High
		Lung and other cancers	High
		Birth defects incidence	High
		Immunological effects (in relationship to complex mixes)	Medium
		Renal diseases (in relationship to metal and solvents)	Medium
		Immunological and mental development	Medium
Drinking water contamination	Stomach cancer	Nitrates	High
	Colon, rectum, bladder cancer	Byproducts of chlorine disinfection	High
	Cancer, cardiovascular disease, reproductive outcomes	Arsenic	Medium
	Neurological disorders, mental development	Population exposure to lead	Medium

Pollutant	Health outcomes	Priority issues	
Pesticides	Cancer, neurological symptoms, reproductive outcomes	Chemicals contaminating water from hazardous waste sites	High
	Cancer, birth defects, neurological disorders, immunological disorders	Environmental exposure of the general population	High
		Populations that have been affected by acute episodes of intoxication	High
		Development of biological markers of past cumulative exposure	High
		Use of immunological changes as indicators of early response and as markers of exposure	High
Electromagnetic fields	Leukaemia (especially in children)	Mechanisms of action on biological systems	High
	Brain cancer	Definition of proper measure of exposure	High
		Risk associated to residential exposure	High
		Reasons for increased cancer incidence in electrical workers	High
Disasters	Various outcomes	Identification of sensitive populations	High
		Relationship between acute and delayed effects	High
		Biomarkers of exposure	High
		Psychological aspects of both exposure and disease at community level	High
Indoor radon exposure	Lung cancer	Extrapolation of cancer risk estimates from one population to another	Medium
	Leukaemia and other cancers	Role of indoor radon exposure	Medium
	Various outcomes	Development of biomarkers of exposure and early response	Medium
UV radiation	Skin cancer	Role of non-natural sources of exposure (UV and fluorescent lamps)	Medium
	Cataract and other health effects	Genetic susceptibility to UV radiation effects	Medium
		Role of UV radiation	Medium

[a] The individual role of each pollutant as well as that of their interactions and mixtures should be considered for all the issues. Special attention should be given to by-products of the combustion, diesel, petrol and natural gas, particularly in relationship to cancer development.

[b] CO = Carbon monoxide; ETS = Environmental Tobacco smoking.

General Conclusions and Recommendations

In addition to the priorities listed in Table 2, the discussion emphasised issues of a more general nature which should all be taken into account when planning and coordinating studies in environmental epidemiology.

1. International collaboration is often of great value in environmental epidemiology. For example, to cover a large range of exposure and achieve sufficient statistical power, studies from different countries may be combined. Further, international collaboration can also contribute to the harmonisation of methodology. International organisations, particularly WHO, could play a key role in initiating and coordinating these activities.
2. Registers of health effects such as birth defects, cancer, and mortality are useful for environmental epidemiology. It is essential that such registers be maintained and that their quality be continuously assessed. More use should be made of existing registers in environmental epidemiology studies.
3. Studies in environmental epidemiology should never replace or postpone the amelioration of environmental conditions strongly suspected of causing harmful health effects. Epidemiological studies may be used to assess the impact and effectiveness of environmental intervention programs.
4. Adequate funding is crucial in environmental epidemiology research, which often involves the study of weak associations. This generally necessitates large studies and methods that can accurately estimate exposure; however, both tend to be costly. Further, environmental exposures are often greatest in countries where resources are most scarce. International funding agencies should allocate more resources to environmental epidemiology, particularly in the heavily polluted areas in the central and eastern countries of Europe. Other forms of multilateral and bilateral collaboration are also important.
5. In general, training in environmental epidemiology in the WHO European Region is inadequate. It should be strengthened at the undergraduate, postgraduate, and professional levels. International collaboration coordinated by WHO is already taking place, but these activities need additional support.

We believe this meeting was an important opportunity for an exchange of views and opinions among experts in the field of environmental epidemiology to orient research in the field. We hope this effort will be useful for research groups and funding agencies.

ACKNOWLEDGMENTS

The editors wish to thank Dr. Carlo Perucci of the Lazio Region Epidemiology Unit for having given his intellectual and financial support to the organisation and conduct of this meeting. We are also grateful to Dr. Francesco

Forastiere of the Lazio Region Epidemiology Unit, Dr. Pietro Comba of the Italian Higher Institute of Health, and Dr. Roberta Pirastu of the University of Rome for their contribution in the ideation and organisation of the meeting. In addition, our thanks also go to Ms. Manuela Zingales of the WHO European Centre for Environment and Health, Rome Division, and Ms. Paula Carlé and Mrs. Anna Emigli of the Lazio Region Epidemiology Unit for their administrative support before and during the meeting. We are extremely grateful to Ms. Candida Sansone, of the WHO European Centre for Environment and Health, Rome Division, for her invaluable secretarial and administrative work in the organisation, conduct, and follow-up of the meeting as well as in the preparation of this volume.

Roberto Bertollini
Rome, Italy

Michael D. Lebowitz
Tucson, Arizona

Rodolfo Saracci
Lyons, France

David A. Savitz
Chapel Hill, North Carolina

List of Workshop Participants

TEMPORARY ADVISERS

Sir Donald Acheson
Visiting Professor of International
  Health
London School of Hygiene and
  Tropical Medicine
London, United Kingdom
(*Chairperson*)

Dr. Olav Axelson
Department of Occupational Medicine
University Hospital
Linköping, Sweden

Dr. Pier Alberto Bertazzi
Istituto di Medicina del Lavoro
Università degli Studi di Milano
Milan, Italy

Dr. Aaron Blair
Occupational Studies Section
National Cancer Institute
Bethesda, Maryland

Dr. Bert Brunekreef
Department of Epidemiology and
  Public Health
Wageningen Agricultural University
Wageningen, Netherlands

Dr. Kenneth P. Cantor
Environmental Epidemiology Branch
National Cancer Institute
Bethesda, Maryland

Dr. Pietro Comba
Higher Institute of Health
Rome, Italy

Dr. Paul Elliott
Environmental Epidemiology Unit
Small Area Health Statistics Unit
London School of Hygiene and
  Tropical Medicine
London, United Kingdom

Dr. Francesco Forastiere
Lazio Region Epidemiology Unit
Regional Health Authority
Rome, Italy

Dr. Wieslaw Jedrychowski
Department of Epidemiology and
  Preventive Medicine
University Medical School
Cracow, Poland

Dr. Bengt Källén
Department of Embryology
University of Lund
Lund, Sweden

Dr. Michael D. Lebowitz
Respiratory Sciences Center
University of Arizona
College of Medicine
Tucson, Arizona
(*Vice-Chairperson*)

**Dr. Anthony B. Miller, M.B.,
F.R.C.P.**
Department of Preventive Medicine
  and Biostatistics
Faculty of Medicine
University of Toronto
Ontario, Canada

Dr. Igor V. Osechinsky
Head, Department for Population
  Investigations
National Haematological Centre
Moscow, Russian Federation

Dr. Göran Pershagen
National Institute of Environmental
  Medicine
Karolinska Institute
Stockholm, Sweden
(*Rapporteur*)

Dr. Carlo A. Perucci
Director, Lazio Region Epidemiology
  Unit
Regional Health Authority
Rome, Italy

Dr. Roberta Pirastu
Higher Institute of Health
Rome, Italy

Dr. David A. Savitz
University of North Carolina
Chapel Hill, North Carolina

Dr. Lorenzo Simonato
Registro Tumori del Veneto
University of Padua
Padua, Italy

Dr. Benedetto Terracini
Department of Biomedical Sciences
  and Human Oncology
Turin University
Turin, Italy

Dr. Paolo Vineis
Department of Biomedical Sciences
  and Human Oncology
Turin University
Turin, Italy

Dr. David Wegman
Department of Work Environment
College of Engineering
University of Massachusetts
Lowell, Massachusetts

REPRESENTATIVES OF OTHER ORGANIZATIONS

International Agency for Research on Cancer

Dr. Rodolfo Saracci
Lyons, France

OBSERVERS

Dr. Mary M. Agocs
National Institute of Hygiene
Budapest, Hungary

Dr. Marco Marchi
Department of Statistics
University of Florence
Florence, Italy

Dr. Paolo Crosignani
Divisione di Epidemiologia
Istituto dei Tumori
Milan, Italy

Dr. Paolo Paoletti
Istituto di Fisiologia Clinica del CNR
Pisa, Italy

Dr. Riccardo Puntoni
Istituto Nazionale per la Ricerca sul
   Cancro
Genoa, Italy

Dr. Maurizio Di Paola
ENEA-CASACCIA
Rome, Italy

Dr. Adele Seniori Costantini
Occupational Epidemiology Branch
Centro per lo Studio per la
   Prevenzione Oncologia (CSPO)
Florence, Italy

Dr. Carrado Magnani
Servizio di Epidemiologia dei Tumori
Turin, Italy

WORLD HEALTH ORGANIZATION

Regional Office for Europe

Dr. Roberto Bertollini
Epidemiologist
WHO European Centre for
   Environment and Health
Rome Division

Miss Candida Sansone
Secretary
WHO European Centre for
   Environment and Health
Rome Division

Dr. Michal Krzyzanowski
Epidemiologist
WHO European Centre for
   Environment and Health
Bilhoven Division

Miss Manuela Zingales
Secretary
WHO European Centre for
   Environment and Health
Rome Division

Headquarters

Dr. Tord Kjellström
Prevention of Environmental
   Pollution
Division of Environmental Health

Contents

Part 3: Methods

PART 1: EXPOSURE

1

AIR POLLUTION

David H. Wegman

CONTENTS

INTRODUCTION

The primary objective of this chapter is to examine epidemiologic evidence for the effects of outdoor air pollution on human health. The focus is directed to epidemiologic findings relating specific agents to human health and an effort is made to call attention to the areas where further epidemiologic study can be expected to be most rewarding.

A number of limitations occur when applying the range of epidemiologic methods to the study of air pollution and human health. These include the problems presented by: (1) the generally ubiquitous nature of air pollution exposure; (2) the difficulty in trying to describe meaningful variation in exposure within microclimates; (3) the complex range and timing of daily life activities; (4) lack of specificity of many health outcomes; and (5) the many confounding factors which must be accounted for when estimating the actual impact of air pollutants on human health.

The description of the different types of epidemiologic approaches that have been used to study health effects of air pollutants and the problems generic to each of them will be covered separately. Consequently, the remainder of this chapter provides a brief summary discussion of epidemiologic evidence for all the major, and a few minor, air pollutants and some less well-studied agents followed by suggestions for research priorities where epidemiology can be expected to play an important role.

OZONE

Ozone is a highly reactive chemical agent which is considered to be the most important constituent of summer "smog". Levels of ozone pollution are much higher in summer and vary over the course of the day with the highest levels in midafternoon. Ozone is a potent oxidant and highly reactive, especially when in contact with the pulmonary epithelium. Consequently it is recognised as a source of irritation of mucous membranes and of pulmonary epithelial inflammation.

Common symptoms associated with ozone exposures include cough, phlegm, and dyspnea. Less commonly there are reports of substernal pain. Objective changes in pulmonary function have been associated with ozone, most strikingly increased airway reactivity and resistance.

Epidemiologic Studies

Ozone-related acute respiratory health effects have been the primary focus of epidemiologic studies, although attempts have also been made to examine the long-term impact of chronic ozone exposures as well. The acute or short-term studies have used designs which focus on the higher summer levels of ozone pollution as well as on groups which are either potentially more susceptible (asthmatics) or suffering fewer long-term effects (children).

One common design, diary studies, takes account of the episodic nature of ozone exposures as well as the short duration of many of the acute effects. These studies provide subjects with structured diaries in which to record such items as daily activities and symptoms over a period of days to as much as a year or more. Records can then be related to independent measures or estimates of exposure collected simultaneously. Whittemore and Korn[1] developed a new two-stage logistic modelling approach and used it to identify a relationship between asthma attacks recorded in daily diaries and ozone exposures measured by ambient monitoring stations. The great strength of this approach is that it uses to advantage repeated observations of reversible events in study subjects. For example, they reported a 20% increase in asthma attacks with a 0.01 ppm increase in oxidant levels. Their studies raised the importance of measuring the impact of previous health events on future events when the events can reoccur frequently. This work was replicated in Houston, Texas with more detailed estimates of exposure and activity.[2,3]

Related work by Schwartz and Zeger[4] used population health data collected for more general purposes to relate ambient ozone levels to daily symptom reports. They demonstrated an exposure-related increase in frequency of chest discomfort and eye irritation and an increase in duration of cough and phlegm, raising the problem of how best to measure the relative importance (or interaction) of frequency, severity, and duration of reversible health events.

A second design which has been employed by several investigators uses children in controlled environments such as at summer camps or in schools. The controlled environment allows for regular evaluation of symptoms or pulmonary function. In the camps, the children are physically active, out-of-doors, and in rural environments where during summer months an otherwise low pollution environment is affected by high regional ozone levels (along with other pollutants). These studies provide evidence of ozone-related acute lung function reductions larger than seen in chamber studies.[5] The responses were not homogeneous and reasons for this heterogeneity of response need to be sought. The differences between chamber and camp studies also need to be explored in terms of cumulative effects, potentiation by other pollutants, and length of exposure. Studies in schoolchildren also demonstrated reversible changes in pulmonary function related to several different latencies between exposure episodes and symptom onset.[6,7]

In a third design, children and/or adults are followed over several years to measure the effects of ozone on either lung development or lung growth. Detels' work in Los Angeles suggests accelerated ageing of the lung due to chronic exposure to a mix of photooxidants.[8] These studies suffer from two limitations: (1) it is difficult to characterise chronic exposure for individual subjects, and (2) cohort follow-up is seriously limited by population mobility.

A final group of studies has sought to take advantage of general health records such as hospital admissions or local mortality. These have provided some suggestive leads but, for the most part, have not been very informative. Using an ecologic approach, Bates showed correlations of hospital admission rates for respiratory causes with elevated ozone (particularly in association with SO_2) in Ontario, but in British Columbia the associations were with SO_2 and sulphates only.[9,10] Kinney has suggested that total mortality as well as cardiovascular disease mortality increases with pollution levels.[11] In both types of studies, there is difficulty with appropriate assignment of subject exposures, particularly lifetime exposure as might be appropriate for mortality.

SO$_2$, PARTICULATES, AND ACID AEROSOLS

These three important air pollutants are considered together. Since they result from common sources, evaluation of their independent effects on the health of a population is difficult or impossible. Since control options would most likely affect all equally, this problem may not be critical.

These three pollutants are associated most importantly with winter smog. The production of SO$_2$ and particulates is generally a result of burning fossil fuels. Under appropriate atmospheric conditions, the two react to produce acid aerosols. Other sources of atmospheric particulates include pollen and contributions from manufacturing and selected agricultural processes.

Sulfur dioxide, a highly water-soluble irritant acts primarily on the upper airways. The effects of particulates vary by their size and by their composition while acid aerosols cause decreased mucociliary clearance and altered phagocytosis and are associated with increased airway reactivity and resistance.

Epidemiologic Studies

These three agents were the subject of the earliest studies of the dramatic potential harm due to air pollution. Investigations using general health records were employed to demonstrate how the mid-century pollution disasters in London and in Pennsylvania resulted in striking increases in overall mortality. Recent reanalyses of the data from these incidents provide evidence that the mortality excess was related primarily to acid aerosol levels.[12]

More recent studies have examined mortality with respect to daily particulate and SO$_2$ air pollution levels. Schwartz and Dockery found overall mortality as well as mortality from chronic obstructive pulmonary disease (COPD), pneumonia, and cardiovascular disease were related separately to both pollutants.[13] Their multivariate analysis, however, demonstrated an effect due only to particulates when both pollutants were examined in the same model. Conversely, studies in Greece have shown a relationship between mortality and SO$_2$ pollution, but not to the indirect measure of particulates which was used.[14] Most recently, further epidemiologic evidence for an association of particulate <10 μm aerodynamic diameter (PM$_{10}$) exposure with mortality excess[15] has led to a call for more targeted research in order to characterise the mechanism behind this repeatedly noted association.[16]

Bates' studies of hospital admissions (previously described in the discussion of ozone) showed a relationship to sulphate pollution in both Ontario and British Columbia.[9,10] The relationship in Ontario could not be separated from a parallel relationship to oxidant air pollution. Within the recent past, physician visits and hospitalisations have been associated with acid aerosol air pollution episodes in both Great Britain and Germany.[17,18]

As with studies of effects from ozone pollution, diary studies have been effective in identifying ill health related to these pollutants. Schwartz found evidence both for lower respiratory tract symptoms[19] and for emergency room visits for asthma[20] associated with PM_{10} concentrations, while Ostro found both H^+ and $PM_{2.5}$ associated with asthma status, cough, and shortness of breath in a panel of asthmatics.[21]

Studies of pulmonary function have documented acute reversible changes of unknown importance in children experiencing exposure in air pollution episodes measured by SO_2 and by total suspended particulate (TSP).[22,23] When similar studies were carried out in an area of high PM_{10} but low SO_2 (Utah Valley), peak flow was affected along with elevated respiratory symptoms in children with chronic respiratory symptoms.[24]

Chronic respiratory effects have also been noted by Van der Lende when evaluating subjects exposed to high levels of SO_2 and particulates 12 years previously, although now living in reduced SO_2 environments.[25] In the Harvard Six Cities study, Dockery demonstrated a measure of the association of chronic bronchitis, cough, and respiratory illnesses with particulate pollution, but Speizer was able to improve on the association when direct measures of acid aerosols were substituted for the particulate measures.[26,27]

Unlike with ozone, point sources of sulphates and particulates can be studied. Kitagawa examined the pollution pattern associated with elevated asthma occurrences in Yokkaichi.[28] His work suggested that the respiratory illnesses resulted from a pattern of pollution with acid aerosols which were a result of stack effluent from a titanium dioxide plant.

Occupational studies have also contributed to understanding the population effects of SO_2. Several investigators attempted to examine acute and chronic pulmonary function effects of SO_2 levels in smelter workers. However, only Smith has been able to show consistent dose-related effects after taking account of the use of personal protective equipment.[29] He provided evidence of both cross-shift and cross-year changes in FEV_1 (forced expiratory volume in 1 sec) levels in workers exposed below the occupational standard. Smith's unusual approach permitted substantial adjustments in SO_2 dose estimates not possible in any of the other published studies.

NITROGEN OXIDES

The nitrogen oxides occur in outdoor air pollution as a result of automobile emissions from high temperature combustion. The highest ambient measures of NO_x have been in the Los Angeles basin. There is experimental evidence that nitrogen oxides affect the resistance to infection including the immune system.

Epidemiologic Studies

Epidemiologic investigations have focused on study of NO_x effects through measurement of respiratory infections in populations with differential exposures to these agents. Few such studies have been carried out, due in part to the difficulty of adequate measurement and in part to lack of se ing, where exposures to NO_x vary without other air pollutants varying as well.

The best studies of outdoor exposures were carried out by Shy and colleagues in Tennessee where air pollution levels were affected by a local munitions factory. Over several years, childhood respiratory illnesses were examined in several schools with differing NO_2 levels in the local environments.[30-32] These studies used biweekly questionnaires on acute respiratory illnesses as well as physician records of lower respiratory tract infections. Some evidence of associations of these outcomes with NO_2 level was provided but consistency between age groups and among different outcomes was not strong and effects were not consistently related to the full gradient of exposure. Recently NO_2 effects have been studied in indoor environments and these are covered elsewhere.

CARBON MONOXIDE

Exposure to carbon monoxide (CO) shows large temporal and spatial variation because it is produced in any incomplete combustion of organic material. As a component of outdoor air pollution causing general population exposures, the primary source of CO is automobile exhaust and the primary exposure circumstance is while driving or riding in a motor vehicle.

The best described human health risk associated with carbon monoxide exposure is that of acute overexposure resulting in fatal poisoning. There is also evidence that chronic overexposure is associated with increased cardiovascular events (myocardial infarctions [MI]) as well as increase in anginal pain.

Epidemiologic Studies

The examination of risks from CO exposure in the general environment has been difficult. Ecologic studies in the 1970s provide some support that this exposure source can be of some risk. For example, hospital admissions for myocardial infarction in Los Angeles County were associated with CO levels measured at the monitoring stations closest to each of 35 hospitals.[33] Results suggested that the risk of MI was associated with CO elevations >8 ppm and only present during air pollution episodes with elevated CO levels. Kurt found similar results when examining hospital admissions for cardiorespiratory complaints in Denver.[34]

One study of bridge and tunnel workers in New York City took advantage of 30 years of CO measures in toll booths and the tunnels. Risk of mortality from arteriosclerotic heart disease was higher among the tunnel workers and decreased following reduction of exposures from improved tunnel ventilation.[35] These studies suffer the shortcomings of all ecologic studies as well as the failure to account for the effects of other air pollutants.

The hypothesis that environmental exposure to CO was associated with myocardial events was assessed somewhat more directly by a review of hospital records in Baltimore. Ambient CO, COHb levels, and cardiac illnesses were examined and no associations were observed between CO and sudden deaths or admissions for MI.[36] Lambert, however, used a technology-intensive approach to study electrocardiographic evidence of myocardial ischemia associated with environmental CO exposure. Twenty subjects with ischemic disease wearing personal monitors for CO and ambulatory electrocardiograph (ECG) recorders were found to have ST segment depressions on their ECG records that were significantly related to CO levels.[37]

With the exception of one study examining reports of headache, the impact of ambient CO exposure on the neurologic system has not been studied in community-based studies. Schwartz found headache recorded in a daily symptom diary was associated with increased ambient CO levels of >6.5 ppm.[38] Other sources of pollutant due to automobile exhaust, however, need to be controlled to establish a clear association with CO.

Early efforts to examine adverse reproductive outcomes associated with CO exposure have not been successful. For example, an ecologic study in Los Angeles of sudden infant death syndrome (SIDS) and ambient air pollution levels (lagged by two weeks) did not show an association.[39] An ecologic approach was also used in Denver to study birthweight and ambient CO.[40] No association was found but misclassification of CO exposure, along with significant unaccounted for other exposures (cigarette smoking), were significant limitations noted by the authors.

Studies of respiratory effects in working populations exposed to high CO levels have suggested adverse effects of CO; however, none has been able to account adequately for simultaneous exposure to the other pollutants, in particular, other products of combustion. At least one major community-based examination of CO effects on the lung did not show associations with decrements in lung function.[41]

LEAD

Although there are many sources of environmental lead contamination, the source which has contributed most to environmental air pollution is that derived from organic lead compounds used as gasoline additives. The association of this use with general environmental lead contamination and elevated blood lead values in the general population is well accepted. Whether the uptake of lead in humans is primarily a result of direct inhalation or through ingestion of water and food products contaminated by lead, however, is not known. Regardless of route of entry, the removal of lead as a gasoline additive in the U.S. has been associated with a well-documented reduction in population blood lead values.

Lead has been associated with a variety of health effects in humans. These include anaemia, hypertension, gastrointestinal distress, adverse reproductive outcomes, and both peripheral and central nervous system abnormalities. Most of the health effects information has been gained from studying disease in workers exposed to lead during employment, exposures much higher than in the general environment. The effects that have been documented at levels low enough to be of general concern are hypertension and central nervous system effects.

Epidemiologic Studies

Needleman[42] has studied intelligence and behaviour in children with low but varied levels of dentine lead. He examined first and second grade children (7–9 years old) using measures of lead in deciduous teeth to estimate lead body burden.[42] The subjects were scored on IQ and evaluated by their teachers on a number of behavioural parameters. A strong inverse association of lead with IQ was demonstrated and negative behavioural scores were shown to increase with dentine lead level. These findings have been confirmed by others.[43] In addition, Needleman followed these children through high school and found that those with evidence of elevated dentine lead in primary school were more likely to leave school prematurely and to have a reading disability.[44]

Studies of lead's impact on blood pressure have also suggested effects at low levels. Harlan examined patients with hypertension and found systolic and diastolic blood pressure associated with blood lead levels in men over a wide age range. Pirkle, studying these data further, found the association was significant at blood lead levels which ranged between 7 and 34 µg/dl.[45] Data from the second U.S. NHANES survey were also studied using the carefully standardised measures of blood pressure and relating them to blood lead levels in participants from the general population.[46] The results were similar to those found in patients having documented hypertension with a linear association between blood level and both systolic and diastolic blood pressure.

CARCINOGENS

There is a general concern that ambient air contains a variety of agents in sufficient concentration to influence cancer risk for the general population. It is extremely difficult, however, to document this risk due both to the difficulty in measurement of exposure to the low level of carcinogens and the fact that cancer is a "rare" disease. The primary pollutants that have been considered candidates as important human carcinogens in ambient air pollution include asbestos, benzene, and diesel exhaust; and recently it has been suggested that acid aerosols should be added to the list.

Asbestos

General population exposure to asbestos materials result from two major sources: spray insulation material used in building construction between approximately 1950 and 1975 to provide protection against fire, and brake and clutch linings used in motor vehicles. Exposure to the former occurs largely indoors, but the latter does result in increased outdoor sources of ambient asbestos fiber exposures. Use of brakes and the clutch in motor vehicles result in abrasion and fragmentation of the asbestos contained in the linings. This results in fibers released to the general atmosphere, a circumstance which largely explains the tenfold difference in measures of ambient fibers in urban and rural air.[47] Although a minor source for general population exposure, local elevations in exposure have been noted in rural areas near asbestos-containing outcroppings, in areas near demolition or use of sprayed asbestos insulation, and/or near areas where concentrated high volume vehicle braking occurs.

Asbestos is accepted as a human carcinogen causing lung cancer and mesothelioma and also is associated with fibrotic lung disease (asbestosis). These risks have been characterised and quantified in working populations, while evidence of risk in the general population is limited to case reports among family members of asbestos workers or to those who live in the vicinity of mines or manufacturing plants processing or using asbestos materials.

Benzene

General population exposure to benzene is primarily associated with four sources: cigarette smoking (active as well as passive); home use of solvents (generally with less than 0.1% benzene) or gasoline (with up to 1–2% benzene); leaky underground storage tanks (largely resulting in contamination of water supplies); and automotive emissions from the fuel chain (especially the carburetor) or during the process of refueling. In the U.S., the EPA has estimated, however, that general population exposures to benzene result from indoor sources to a far greater extent than from outdoor sources.[48]

Benzene has been accepted as a human leukemogen as well as a cause of aplastic anaemia. Identification of this risk, as with asbestos, has been documented in working populations. Risk in the general population is estimated only by extrapolation from these occupational studies.

Diesel Exhaust

Exposure to diesel exhaust is common, resulting from the substantial use of diesel engines to power larger motor vehicles and railroad trains. In the U.S., diesel-powered engines are used in automobiles and lightweight trucks in less than 3% of vehicles. However, in trucks and buses, they have been more widely adopted so that 20% of the lighter vehicles use diesel engines while as high as 90% of the heaviest do so. No estimates of general air pollution from diesel exhaust exist. Diesel soot, the most commonly used marker for diesel exhaust, is difficult to distinguish from small carbonaceous particles resulting from other sources of combustion. Search for a more specific marker has, as yet, been unsuccessful.

Concern with diesel exhaust derives from several findings.[49] It has been shown to be a mutagen in short-term bacterial assays and has been shown to be a pulmonary carcinogen in studies of rats. Studies in other animal systems have been more mixed (in mice) or negative (in hamsters). Occupational studies in a variety of settings have been reasonably consistent in showing excess lung cancer. The level of excess risk identified in these studies, however, is on the order of 20–30%, at about the limit of detection for epidemiologic studies. It has been suggested that some combination of size-selective sampling and physical chemical analysis will probably be required. A unique biomarker has also been sought, but without success.

Acid Aerosols

In an extensive review by Soskolne, it has been proposed that acidic pollution contributes to human cancer risk.[50] In 1992, the International Agency for Research on Cancer (IARC) reported that there was sufficient evidence to support identifying inorganic acids as human carcinogens.[51]

PRIORITY RESEARCH NEEDS

The challenges to improving knowledge about the above pollutants through further epidemiologic study are often uniquely related both to the specific characteristics of the pollutant and to the special circumstances of the setting to be studied. Yet, there are a number of common threads that connect the efforts to use the epidemiologic method to study the different agents reviewed. An attempt is made to call attention to those issues that are most likely to benefit future epidemiologic studies or where the epidemiologic approach is likely to play an important role in expanding understanding of the health burden due to air pollution.

1. The studies of ozone exposure to children in summer camps emphasize the advantages that come from conducting studies in locations where elevated pollutant exposures can be predicted sufficiently well to examine short-term effects. Studies in these settings should be repeated over several cycles (summers) to determine the effect of intermittent excessive exposures on lung growth and on development of irreversible lung abnormalities. A related type of environment is where a point source allows a better controlled study of a general air pollutant. Opportunities such as the studies near the munitions or the titanium dioxide plants are examples.

2. Diary studies have provided a great deal of information on several pollutants. The greatest effort has been directed at respiratory symptoms but diaries could be used to examine pollutant effects on headache and cardiovascular symptoms as well. Here, as in other study forms, there is need to take advantage of the expanding developments in psychophysics to allow improvements in questions about symptom type and to provide symptom intensity scales in a semi-quantitative form. These should allow more useful comparisons within and across populations that differ in geography, in ethnic makeup, in age, and especially in type of pollution exposure.

3. Since many symptoms studied are reversible, a component of diary studies should be directed to examining the interrelationships of symptom frequency, intensity, and duration; and the importance of recovery or refractory periods with respect to frequently recurring effects.

4. In many respects, the prospective cohort study is ideal to study pollutant effects; but often these studies are limited by losses to follow-up through population mobility. Settings should be sought where air pollution exposures can be well characterised and where circumstances provide for high probability that population mobility will be low. Eastern Europe may provide such opportunities. New ways to integrate study designs that incorporate advantages of prospective cohort studies (both dynamic and fixed) with those of cross-sectional studies should also be sought.

5. By far the most difficult problem in epidemiologic studies of air pollutants is inadequate measurement of personal exposure. Correcting this problem for epidemiologic purposes requires development of practical and accurate personal exposure monitors along with convenient and acceptable biomarkers. These developments will require attention to real-time monitors as well as time-integrated ones.

6. The problems of measuring exposure and dose in epidemiologic studies include: (1) appropriate models to estimate dose and cumulative dose, (2) improvements in studying the effects of latency on the relationship between exposure and the timing on effects, (3) methods to account for the combined effects of several agents affecting a common end-point, and (4) examination of variability in exposure levels to one or several agents (within, as well as across days and seasons) to determine whether these have dominant or relatively unimportant effects on dose.

7. More advantage should be taken of settings in which major impacts are being made on control of air pollutants. One could then examine a well-characterised subset of the population (e.g., those with history of asthma or frequent childhood respiratory illnesses) and follow them as the pollution

burden is reduced to determine whether health status improves and, if so, in what way.

8. A number of factors which can influence or modify dose-to-target tissues have not yet been considered in studies of air pollutants. Important among these are the need to find effective ways to monitor level of physical activity within and between days and to document the nature of avoidance behaviours associated with irritant exposures which alter generally assumed patterns of behaviour. Better characterisation of these could be important in measuring the impact of the major air pollutants.

9. The role of infection needs to be better characterised for nitrogen oxide exposures but attention needs to be directed to the importance of infections for other pollutants as well. Study designs need to account both for infections as a source of change in susceptibility and increased infections as a consequence of certain types of pollution or pollution patterns. Diary studies should be helpful here, but a method to establish a reasonable retrospective history will need to be developed as well. Such studies will also need to be carried out over longer periods of time than have been used to date.

10. Biomonitoring methods can serve an important role in certain types of air pollution studies. Benzene, lead, and carbon monoxide already have excellent means for monitoring exposure that are noninvasive. These should be better utilised in future studies but probably require some improvements to make them manageable in less well-supervised field settings.

11. There is little hope that epidemiologic studies of airborne carcinogens in general population studies can be developed to directly estimate cause-specific cancer risk in the general population. It is much more likely the case that the risks will need to be extrapolated from studies done in more highly exposed groups. Methods therefore need to be developed to improve the means of extrapolating risk to subjects not studied or able to be studied in the higher exposures (the young, the ill, or the aged). Means to verify the validity of such extrapolations may be the most important focus for improving "understanding" of air pollution and cancer.

12. Although working populations are not representative of the general population, there are numerous advantages of occupational settings for well-designed and controlled epidemiologic investigations. Therefore, opportunities should be sought to undertake epidemiologic studies in occupational settings wherever the studies can be designed to meet objectives of both air pollution and occupational exposure risks.

REFERENCES

1. Whittemore, A. S. and Korn, E. L., Asthma and air pollution in the Los Angeles area. *Am. J. Public Health* (1980) 70:687-96.
2. Holguin, A. H., Buffler, P. A., Contant, C. F., Stock, T. H., Hotchmar, D., Hsi, B. P., Jenkins, D. E., Gehan, B. M., Noel, L. M., and Mei, M., The effects of ozone on asthmatics in the Houston area. *Trans. Air. Pollut. Control Assoc.* (1985) TR-4:262-280.

3. Contant, C. F., Stock, T. H., Buffler, P. A., Holguin, A. H., Hehan, B. M., and Kotchmar, D. J., The estimation of personal exposures to air pollutants for a community-based study of health effects in asthmatics — exposure model. *J. Air Pollut. Control Assoc.* (1987) 37:587-594.

4. Schwartz, J. and Zeger, S., Passive smoking, air pollution, and acute respiratory symptoms in a diary study of student nurses. *Am. Rev. Respir. Dis.* (1990) 141:62-67.

5. Spektor, D. R., Mao, J., He, D., Thurston, G. D., Hayes, C., and Lippmann, M., Health effects of ambient ozone on healthy children at a summer camp. *Am. Rev. Respir. Dis.* (1991) 141:A71.

6. Kinney, P. L., Ware, J. H., Spengler, J. D., Dockery, D. W., Speizer, F. E., and Ferris, B. G., Jr., Short-term pulmonary function change in association with ozone levels. *Am. Rev. Respir. Dis.* (1989) 139:56-61.

7. Castillejos, M., Gold, D. R., Dockery, D., Tosteson, T., Baum, T., and Speizer, F. E., Effects of ambient ozone on respiratory function and symptoms in Mexico City schoolchildren. *Am. Rev. Respir. Dis.* (1992) 145:276-282.

8. Detels, R., Tashkin, D. P., Sayre, J. W., Rokaw, S. N., Massey, F. J., Coulson, A. H., and Wegman, D. H., The UCLA population studies of chronic obstructive pulmonary disease: IX. Lung function changes associated with chronic exposure to photochemical oxidants a cohort study among never smokers. *Chest* (1987) 92:594-603.

9. Bates, D. V. and Sizto, R., Air pollution and hospital admissions in southern Ontario. *Envir Res.* (1987) 43:31-331.

10. Bates, D. V., Baker-Anderson, M., and Sizto, R., Asthma attack periodicity: a study of hospital emergency visits in Vancouver. *Environ. Res.* (1990) 51:51-70.

11. Kinney, P. L. and Ozhaynak, H., Associations of daily mortality and air pollution in Los Angeles County. *Environ. Res.* (1991) 54:99-120.

12. Thurston, G. D., Ito, K., Lippmann, M. and Hayes, C., Reexamination of London, England, mortality in relation to acid aerosols during 1963-72 winters. *Environ. Health Perspect.* (1989) 79:73-82.

13. Schwartz, J. and Dockery, D.W., Particulate air pollution and daily mortality in Steubenville, Ohio. *Am. J. Epidemiol.* (992) 135:12-9.

14. Hatzakis, A., Katsouyanni, K., Kaolandidi, A., Day, N., and Trichopoulos, D., Short-term effects of air pollution on mortality in Athens. *Int. J. Epidemiol.* (1986)15:73-81.

15. Pope, C. A., III, and Kanner, R. E., Acute effects of PM_{10} pollution on pulmonary function of smokers with mild to moderate COPD. *Am. Rev. Respir. Dis.* (1993) 147:1336-1340.

16. Utell, M. J. and Samet, J.M., Particulate air pollution and health. (Editorial). *Am. Rev. Respir. Dis.* (1993) 147:1334-1335.

17. Ayres, J., Fleming, D., Williams, M., and Mcinnes, G., Measurement of respiratory morbidity in general practice in the United Kingdom during the acid transport event of January, 1985. *Environ. Health Perspect* (1989) 79:83-88.

18. Wichmann, H. E., Mueller, W., Alhoff, P., Beckmann, M., Bocter, N., Csicsaky, M. J., Jung, M., Molik, B., and Schoenberg, G., Health effects during a smog episode in West Germany in 1985. *Environ. Health. Perspect.* (1989) 79:89-99.

19. Schwartz, J., Dockery, D. W., Ware, J. H., Spengler, J. D., Sypig, D., Koutrakis, P., Speizer, F. E., and Ferris, B. G., Jr., Acute effects of acid aerosols on respiratory symptoms reporting in children. Presented at the 82nd Annual Meeting of the Air Pollution Control Association, (1989) Paper no 89-92.1.

20. Schwartz, J., Slater, D., Larson, T. V., Pierson, W. E., and Koenig, J. Q., Particulate air pollution and hospital emergency room visits for asthma in Seattle. *Am. Rev. Respir. Dis.* (1993) 147:826-831.

21. Ostro, B. D., Lipsett, M. J., Wiener, M. B., and Selner, J. C., Asthmatic responses to airborne acid aerosols. *Am. J. Public Health* (1991) 81:694-702.

22. Dassen, W., Brunekreef, B., Hoek, G., Hofschreuder, P., Staatsen, B., de Groot, H., Schoutan, E., and Biersteker, K., Decline in children's pulmonary function during an air pollution episode. *J. Air Pollut. Control Assoc.* (1986) 36:1123-1127.

23. Dockery, D. W., Ware, J. H., Ferris, B. G., Jr., Speizer, F. E., Cook, N. R., and Herman, S. M., Change in pulmonary function associated with air pollution episodes. *J. Air Pollut. Control Assoc.* (1982) 32:937-942.

24. Pope, C. A., III, Dockery, D. W., Spengler, J. D., and Raiseene, M. E., Respiratory health and PM_{10} pollution. *Am. Rev. Respir. Dis.* (1991) 144:668-674.

25. van der Lende, R., Kok, T. J., Reig, R. P., Quanger, P. H., Schouten, J. P., and Orie, N. G. M., Decreases in VC and FEV_1 with time: indicators for effects of smoking and air pollution. *Bull. Eur. Physiopathol. Respir.* (1981) 17:775-792.

26. Dockery, D. W., Speizer, F. E., Stram, D. O., Ware, J. H., and Spengler, J. D., Effects of inhalable particles on respiratory health of children. *Am. Rev. Respir. Dis.* (1989) 139:587-594.

27. Speizer, F. E., Studies of acid aerosols in six cities and in a new multi-city investigation: design issues. *Environ. Health Perspect.* (1989) 79:61-67.

28. Kitagawa, T., Cause analysis of the Yokkaichi asthma episode in Japan. *J Air Pol. Control Assoc.* (1984) 34:743-746.

29. Smith, T. J., Peters, J. M., Reading, J. C., and Castle, C. H., Pulmonary impairment from chronic exposure to sulphur dioxide in a smelter. *Am. Rev. Respir. Dis.* 1977; 116:31-39.

30. Shy, C. M., Creason, J. P., Pearlman, E., McClain, K. E., Benson, F. B., and Young, B. B., The Chattanooga school study; effects of community exposure to nitrogen dioxide. II. Incidence of acute respiratory illness. *J. Air Pollut. Conttrol Assoc.* (1970) 20:582-588.

31. Pearlman, M. E., Finklea, J. P., Creason, J. P., Shy, C. M., Young, M. M., and Horton, R. J. M., Nitrogen dioxide and lower respiratory illness. *Pediatrics* (1971) 47:391-398.

32. Love, G. J., Lan, S. P., Shy, C. M., and Riggan, W. B., Acute respiratory illness in families exposed to nitrogen dioxide ambient air pollution in Chattanooga, Tennessee. *Arch. Environ. Health* (1982) 37:75-80.

33. Cohen, S. I., Deane, M., and Goldsmith, J. R., Carbon monoxide and survival from myocardial infarction. *Arch. Environ. Health* (1969) 19:510-517.

34. Kurt, T. L., Mogielnicki, R. P., Chandler, J. E., and Hirst, K., Ambient carbon monoxide levels and acute cardiorespiratory complains: an exploratory study. *Am. J. Public Health* (1979) 69:360-363.

35. Stern, F. B., Halperin, W. E., Hornung, R. W., Ringenburg, V. L., and McCammon, C. S., Heart disease mortality among bridge and tunnel officers exposed to carbon monoxide. *Am. J. Epidemiol.* (1988) 128:1276-1988.

36. Kuller, L. H., Radford, E. P., Swift, D., Perper, J. A., and Fisher, R., Carbon monoxide and heart attacks. *Arch. Environ. Health* (1975) 30:477-482.

37. Lambert, W. E., Cardiac Responses to CO in the Community Setting. Dissertation. University of California, Irvine, 1992.

38. Schwartz, J. and Zeger, S., Passive smoking, air pollution, and acute respiratory symptoms in a diary study of student nurses. *Am. Rev. Respir. Dis.* (1990) 141:62-67.

39. Hoppenbrowers, T., Calub, M., Arakawa, K., and Hodgmen, J. E., Seasonal relationship of sudden infant death syndrome and environmental pollutants. *Am. J. Epidemiol.* (1981) 113:623-635.

40. Alderman, B. W., Baron, A. E., and Savitz, D. A., Maternal exposure to a neighborhood carbon monoxide and risk of low birthweight. *Pub. Health Rep.* (1987) 102:410-414.

41. Lebowitz, M. D., Collins, L., and Holberg, C. J., Time series analysis of respiratory responses to indoor and outdoor environmental phenomena. *Environ. Res.* (1987) 43:332-341.

42. Needleman, H. L., Gunnoe, C., Leviton, A., Reed, R., Peresie, H., Maher, C., and Barrett, P., Deficits in psychological and classroom performance of children with elevated dentine lead levels. *N. Engl. J. Med.* (1979) 300:689-695.

43. Needleman, H. L. and Gatsonis, C.A., Low-level lead exposure and the IQ of children: a meta-analysis of modern studies. *J. Am. Med. Assoc.* (1990) 263:673-678.

44. Needleman, H. L., Schell, A., Bellinger, D., Leviton, A., and Allred, E. N., The long term effects of exposure to low doses of lead in childhood: an 11-year follow-up report. *N. Engl. J. Med.* (1990) 322:83-88.

45. Pirkle, J. L., Schwartz, J., Landis, J. R., and Harlan, W. R., The relationship between blood lead levels and blood pressure and its cardiovascular risk implications. *Am. J. Epidemiol.* (1985) 121:246-258.

46. Schwartz, J. and Otto, D., Blood lead, hearing thresholds, and neurobehavioral development in children and youth. *Arch. Environ. Health* (1987) 42:152-60.

47. Lippmann, M., Asbestos and other mineral fibers, in *Environmental Toxicants*, Lippmann, M., Ed., Van Nostrand Reinhold, New York, 1992, chap 22.

48. Goldstein, B. D., and Witz, G., Benzene, in *Environmental Toxicants*, Lippmann, M., Ed., Van Nostrand Reinhold, New York, 1992, chap 15.

49. Mauderlky, J. L., Diesel exhaust, in *Environmental Toxicants*, Lippmann, M., Ed., Van Nostrand Reinhold, New York, 1992, chap 29.

50. Soskolne, C. L., Pagano, G., Cipollaro, M., Beaumont, J. J., and Ciodano, G. G., Epidemiologic and toxicologic evidence for chronic health effects and the underlying biologic mechanisms involved in sublethal exposures to acidic pollutants. *Arch. Environ. Health* (1989) 44:180-191.

51. International Agency for Research on Cancer (IARC). Occupational exposures to mists and vapours from strong inorganic acids and other industrial chemicals, Vol. 54, IARC Monographs on the Evaluation of Carcinogenic Risks to Humans, Lyon, France, (1992).

2 RADON

Olav Axelson

CONTENTS

EPIDEMIOLOGIC EVIDENCE OF HEALTH EFFECTS

It has been known for many decades that radon, a naturally occurring radioactive noble gas, can reach high concentrations in uranium and other mines. There is also convincing evidence that exposure to radon and its decay products can cause lung cancer among miners. More recently, it has been discovered that high levels of radon may occur in dwellings as well. The health effects are less clear in this respect, but several epidemiologic studies on lung cancer and indoor radon have been conducted since the late 1970s. These studies, which are mainly of the case-control type, have indicated some health hazards, but further studies are certainly warranted. The methodological problems encountered in studying the health effects of indoor radon relate especially to the assessment of exposure over long periods of time. Through the studies conducted so far, good progress has been made in this respect, however.

Radon and Decay Products

The radioactive decay of uranium through radium leads to the release of radon, or more precisely the isotope radon-222, which further decays to a series of radioactive isotopes of polonium, bismuth, and lead. The first four of these isotopes are referred to as short-lived radon progeny (or radon daughters) and have half-lives from less than a millisecond up to almost 27 min. There is also another decay chain from thorium through radon-220, or thoron, but the elements in this series are very short-lived and usually of less hygienic concern.

Like radon-222 itself, the decay products polonium-218 and polonium-214 also emit alpha particles. Since the decay products get electrically charged when created, they tend to attach to surfaces and dust particles in the air but some also remain unattached. Consequently, the unattached fraction tends to decrease in dusty air, which is of some importance, since the unattached progeny usually is held responsible for the major part of the alpha irradiation to the bronchial epithelium. The contribution of alpha irradiation from the radon gas itself is thought to be marginal, however.

Although the alpha particles travel less than 100 μm into the tissue, their high energy causes an intense local ionisation, damaging the tissue with a subsequent risk for cancer development. There is also beta- and gamma radiation from some of the decay products of radon, but since these types of radiation have a much lower energy content than the alpha radiation, the effect in this respect tends to be marginal.

Measuring Exposure

To assess the risk of lung cancer for miners, the working level (WL) unit was introduced as a measure of exposure to radon progeny (Holaday, 1955). One working level is any combination of short-lived radon progeny in 1 liter of air that ultimately will release 1.3×10^5 MeV of alpha energy by decay through polonium-214. This amount of radon progeny may also be taken as equivalent to 3700 Bq/m^3 equilibrium equivalent radon (EER) (ICRP 1976) or 2.08×10^{-5} J/m^3. The accumulated exposure to radiation is expressed in terms of working level months (WLM), the month in this context corresponding to 170 h of exposure. The SI unit for exposure is the joule-hour per cubic meter, and 1 WLM is equal to 3.6×10^{-3} Jh/m^3 and may also be taken as 72 Bq-years/m^3.

In mines, the exposure levels have usually been based on direct measurements of the alpha radiation of radon progeny as collected by some filter method and such methods have also been used in dwellings (Kusnetz, 1956).

Several other methods are available for indoor measurements, e.g., film strips sensitive to the alpha radiation from deposited short-lived decay products in which the tracks can be made visible through an etching process. More recently there has been a development towards measurements of radon gas, which is permitted to leak into a capsule and decay to daughter products. These again make tracks on a film strip. Also thermoluminescence dosimeters have come into use for indoor measurements (for further overview see, e.g., IARC, 1988 as well as Svensson et al., 1988). The relation between the concentrations of radon and radon progeny depends on ventilation; usually the radon progeny concentration (in becquerel per cubic meter EER) is taken as half of that of radon.

Radon in Mines and Lung Cancer

Uranium in rock is the source of the high radon levels in uranium mines. There are trace amounts of uranium in many kinds of minerals, however, so that a high radon emanation from rock can occur also in other types of mines, such as iron and zinc-lead mines. Some radon is solved in ground water; and if the mines are wet, radon concentrations tend to increase. Occupational exposures have been high in the past, especially in uranium mines, which may be illustrated by mentioning the need for an exposure category of 3720 WLM and above in one study of uranium miners (Lundin et al., 1971). The exposures to radon progeny in many nonuranium mines have also been relatively high as illustrated in Table 1.

One of the first observations of occupational cancer concerned lung cancer in miners and was first reported in 1879 from Schneeberg in eastern Germany (Härting and Hesse, 1879). Some decades later a similar excess of lung cancer was reported from Joachimsthal in Czechoslovakia (Arnstein, 1913). It was not until the 1950s and 1960s, however, that the etiological role of radon and its decay products was more fully understood and agreed upon. Since the early 1960s, many mining populations with exposure to radon and its decay products have been studied both by the cohort and the case-control techniques as summarised in Tables 1 and 2.

A number of agents other than radon and progeny may be thought of as possibly contributing to miners' lung cancer risk. Miners with a low exposure to radon and radon progeny have shown little or no excess of lung cancer, that is, in coal (IARC, 1988) and potash mining (Waxweiler et al., 1973) and also in iron mining (Lawler et al., 1985).

Indoor Radon

Radon levels in homes depend on building material and, more important, on leakage from the ground (Åkerblom and Wilson, 1982). There is a consid-erable local variation in this radon emanation, even within a few meters, due

Table 1 Main Results of Some Cohort Studies of Lung Cancer in Miners Exposed to Radon and Radon Progeny

Type of Mining; Country	Exposure or Concentration (Means)	Person-Years	Lung Cancer Deaths			Reference; (First Author and Year)
			Observed	Expected	SMR[a]	
Metal, U.S.	0.05–0.40 WL	23,862	47	16.1	2.92	Wagoner, 1963
Uranium, U.S.	821 WLM	62,556	185	38.4	4.82	Waxweiler, 1981; Lundin, 1971
Uranium, Czechoslovakia	289 WLM	56,955	211	42.7	4.96	Placek, 1983; Kunz, 1978
Tin, U.K.	1.2–3.4 WL	27,631	28	13.27	2.11	Fox, 1981
Iron, Sweden	0.5 WL	10,230	28	6.79	4.12	Jörgensen, 1984
Iron, Sweden	81.4 WLM	24,083	50	12.8	3.90	Radford, 1984
Fluorspar, Canada	Up to 2,040 WLM	37,730	104	24.38	4.27	Morrison, 1985
Uranium, Canada	40–90 WLM	202,795	82	56.9	1.44	Muller, 1985
Uranium, Canada	17 WLM	118,341	65	34.24	1.90	Howe, 1986
Iron, U.K.	0.02–3.2 WL	17,156	39	25.50	1.53	Kinlen, 1988
Pyrite, Italy	0.12–0.36 WL	29,577	47	35.6	1.31	Battista, 1988
Uranium nonsmokers, U.S.	720 WLM	7,861	14	1.1	12.70	Roscoe, 1989
Tin, U.K.	10 WLM a year for 30 years	—	15	3.4	4.47	Hodgson, 1990

[a] SMR = standardised mortality ratio.

Table 2 Main Results of Some Case-Control Studies of Lung Cancer in Miners Exposed to Radon and Radon Progeny

Type of Mining; Country	Exposure or Concentration (Means)	Number of Cases to Controls	Number of Exposed Cases	Rate Ratio (max.)	Reference (First Author and Year)
Zinc-lead, Sweden	1 WL	29/174	21	16.4	Axelson, 1978
Iron, Sweden	0.1–2.0 WL	604/(467 × 2 + 137)	20	7.3	Damber, 1982
Iron, Sweden	0.3–1.0 WL	38/403	33	11.5	Edling, 1983
Uranium, U.S.	30–2,698 WLM	32/64	23	Infinite	Samet, 1984
Uranium, U.S.	472 WLM in cases	65/230 (nested)	All	1.5% per WLM	Samet, 1989
Tin, China	515 WLM in cases	107/107	7	(20.0)	Qiao, 1989
Tin, China	373 WLM	74/74	5	(13.2) 1.7% per WLM	Lubin, 1990 (subset of Qiao, 1989)

to cracks and porosity of the ground. Air pressure, temperature, wind conditions, and also various behavioural factors, which influence ventilation, play a role for the concentrations that may build up indoors.

Indoor radon was first measured in Swedish dwellings in the 1950s (Hultqvist, 1956), and levels were found in the range of 20–69 Bq/m3. In the 1980s, measurements of indoor radon in Swedish homes revealed higher levels, i.e., 122 Bq/m3 as an average in detached houses and 85 Bq/m3 in apartments. These concentrations may suggest a general increase of indoor radon over time, but such great variations as from 11 to 3300 Bq/m^3 were observed (Swedjemark, 1987). Levels in the range of 40–100 Bq/m3 have been reported as an average for dwellings in many countries, e.g., U.S. (Nero et al., 1986), Norway (Stranden, 1987), Finland (Castren, 1987), former Federal Republic of Germany (Schmier and Wick, 1985), etc. Considerably higher levels, such as 2000 or 3000, Bq/m3 may occur in many houses, and this is about double the occupational standard for mines in most countries (about 1100 Bq/m3 or 0.3 WL). The recent efforts to improve insulation and preserve energy may have worsened the situation (Dickson, 1978; Stranden, 1979; McGregor et al., 1980).

Studies of Lung Cancer and Indoor Radon

A problem in epidemiology is always to obtain accurate information on exposure, indeed a characteristic feature of environmental epidemiology. Estimates of cumulated exposure to indoor radon tend to be uncertain but might be based on the type of house, on geological features, and particularly on measurements of current concentrations. In order to improve precision in the assessment of individual exposure, some studies have restricted the study population to be rural and longtime residents on the same address.

In correlation studies, the geological characteristics of a region have been considered for indirect assessment of indoor radon exposure. For example, volcanic and sedimentary areas, differing in radioactivity, have been contrasted to create an exposure gradient (Forastiere et al., 1985). Similarly, the occurrence of granite with an increased radioactivity (Archer, 1987), and even worked phosphate deposits with an increased radioactivity (Fleischer, 1981), have been utilised to create a contrast in exposure. Estimated averages of background gamma radiation levels per county have also been used as a correlate with indoor radon (Edling et al., 1982). Examples of some other indicators of radon exposure in correlation studies are data on Ra-226 in water (Bean et al., 1982) or measured levels of radon in water and indoor air (Hess et al., 1983).

As of early 1993, about a dozen case-control studies on indoor radon and lung cancer had been published along with two cohort studies. Most of these studies have been rather limited in size, and the first few were of a pilot character. Almost all of them have shown some effect of indoor radon, usually with an odds ratio of about 2, but the estimates have been higher for certain subgroups. The results of these various studies are summarised in Table 3. The correlation studies referred to have also shown a more or less clear association between lung cancer and indirectly assessed exposure to radon, except in some studies from Canada (Letourneau et al., 1983), China (Hofmann et al., 1985), and France (Dousset and Jammet, 1985).

Table 3   Main Results of Case-Control Studies of Lung Cancer and Exposure to Indoor Radon and Progeny

First Author and Year	Cases to Control	Rate Ratio	Remarks
Axelson, 1979	37/178	1.8	Significant trend; crude exposure assessment
Lanes, 1982	50/50	2.1	Published abstract only
Edling, 1984	23/202	Up to 4.3	Special account for geology and Rn emanation
Pershagen, 1984	Two sets of 30/30		Significantly high exposure for smoking cases
Damber, 1987	604/(467 × 2 + 137)	Up to 2.0	For more than 20 years in wooden houses
Svensson, 1987	292/584	2.2	Women with oat cell cancer
Lees, 1987	27/49	Up to 11.9	Risk of 11.9 for 10 WLM
Axelson, 1988	177/673	Up to 2.0	Clear effect for rural residents only
Svensson, 1989	210/209 +191 (hospital)	Up to 1.8 (in middle exposure category)	Females; rate ratio 3.1 for small cell cancers
Schoenberg, 1990	433/402	Up to 7.2	Females; little effect but in highest exposure category
Blot, 1990	308/356	Up to 1.7 (small cell)	Females in high risk area (China); slight effect for small cell cancer, only
Pershagen, 1992 (= Svensson, 1989 + improved exposure)	210/209 + 191 (hospital)	1.7 as average	Rate ratio up to 21.7 for smokers and high exposure

Note: In addition, two cohorts: Simpson and Comstock, 1983, with some effect (?) and Klotz et al., 1989, with risk ratio of 1.7 (CI95 0.8–3.2).

Other Cancers and Background Radiation, Including Radon

Some of the correlation studies performed have also indicated possible relations for malignancies other than lung cancer, e.g., for pancreatic cancer and male leukaemia (Edling et al., 1982), bladder and breast cancer (Bean et al., 1982), for reproductive cancer in males, and for all cancers taken together (Hess et al., 1983). A high mortality rate from stomach cancer has been reported from New Mexico in an area with uranium deposits (Wilkinson, 1985).

Further reports have followed, suggesting a relationship between radon exposure and leukaemia (Lucie, 1989); and another study could positively correlate the incidence of myeloid leukaemia, cancer of the kidney, melanoma, and certain childhood cancers with average indoor radon exposure in a number of countries (Henshaw et al., 1990). Somewhat surprisingly, lung cancer did not show any significant correlation with radon exposure, as would have been expected.

Inhaled radon reaching the fat cells in the bone marrow in contrast to the radon progeny deposited in the lungs, was thought to be responsible for the induction of myeloid leukaemia (through its further decay). This view has also been opposed (Mole, 1990), but it seems clear that alpha radiation can cause acute myeloid leukaemia, e.g., in Thorotrast patients (BEIR IV, 1988). The filtering of radon progeny through the kidney and the accumulation of radon decay products on the skin were suggested to explain the other correlations seen. Also prostate cancer has shown a correlation with radon exposure (Eatough and Henshaw, 1990) as well as acute lymphoblastic leukaemia in a study from England and Wales (Alexander et al., 1990).

These observations have been criticised (for example, by Butland et al., 1990), but there are also some supportive data from a study in the Viterbo province, Italy. In this province, individuals representing 13 cancer forms and their control subjects were compared with regard to place of residence in areas with different radon levels. This case-control comparison showed increased odds ratios between 2 and 3 for kidney cancer as well as for melanoma and myeloid leukaemia with a suggestive trend for the two forms mentioned first (Forastiere et al., 1992). For lung cancer, there was only a slightly increased risk in the intermediate exposure category, however.

Also two case-referent (case-control) studies on acute myeloid leukaemia might be of interest in this context (Flodin et al., 1981; Flodin et al., 1990). Various exposures were assessed by questionnaires, e.g., occupational exposures, leisure time activities, smoking, and medical care (including use of drugs, X-ray treatment, and X-ray examinations). The type of residence and workplace were considered in terms of potential gamma radiation, since a higher dose would be obtained from concrete and other stone building materials than from wood. Among other results a dose-response relationship was associated with an index for background gamma radiation. However, since increased indoor gamma radiation from stone building materials also means a potential for radon emanation, there is some possibility that radon exposure could have played some etiological role.

Considering exposure to radon and cancers other than of the lung, a novel report from uranium mines in West Bohemia provide important information as far as miners are concerned (Tomasek et al., 1993). Site-specific cancer mortality was studied in 4320 workers for whom an excess of lung cancer was already known from previous observations. There was an increased risk for liver cancer and cancer of the gallbladder and extra hepatic bile ducts; only the latter was closely related to cumulative radon exposure, however.

In view of the findings in other studies, as already referred to, it may be mentioned that malignant melanoma and kidney cancer were also increased among these miners but not significantly so. Stomach cancer was increased for those who started work under the age of 25, but there was no clear relation to radon exposure. Somewhat puzzling results were obtained for haematopoietic cancers. Neither multiple myeloma nor leukaemia and myeloid leukaemia were increased overall, but a statistically significant exposure-response relationship was found between multiple myeloma and cumulative radon exposure. The leukaemia risk (nine cases) was reported as related only to duration of employment in the mine and possibly referable to gamma radiation rather than to radon exposure.

RESEARCH NEEDS

Research on all various aspects of radon has become quite comprehensive, especially with regard to the physical issues. Only environmental epidemiology can finally elucidate the most important aspect, namely, the quantitative impact from indoor radon with regard to cancer risks for the general population, especially for lung cancer.

Mechanisms of Action

The mechanisms through which radon exposure causes damage are mainly a matter of experimental research. Some aspects might be indirectly elucidated by epidemiologic studies, however. First, the absorption and spreading of radon and its decay products to various tissues in the body may be noted (cf. IARC, 1988). As a consequence, there might be carcinogenic effects in organs other than the lung as also indicated by some of the cited studies. Furthermore, chromosomal aberrations and mutations in lymphocytes have been observed in populations exposed to high background radiation, including elevated levels of radon (IARC, 1988; Bridges et al., 1991). The studies do not show fully consistent results, however (Albering et al., 1992).

Because of the various findings referred to as well as the fact that the mechanism behind the health effects of radon and its decay products may involve free-radical formation, an epidemiologic approach could be to compare the excretion of DNA adducts in populations with current high and low exposure to indoor radon. Then the comparisons should involve populations exposed to radon with and without concomitant high background gamma radiation. Quite a sensitive mass screening of damaged DNA fragments in urine like 8-hydroxydeoxyguanosine (8-OHdG) could be a possible approach in this respect, since an automated analytical method has now become available

(Tagesson et al., 1992). This method also seems to be sufficiently cheap for epidemiologic purposes.

Exposure and Disease Assessment

Although measurements of indoor radon are useful for exposure assessment, it is obvious that even extensive measurements in a number of subsequently used residences of an individual would not provide any particularly reliable estimate of his or her cumulated exposure over many decades. The main problem is apparently that most individuals spend considerable time outside the home, either outdoors or in other buildings. Nevertheless, a combination of measurements and judgements with regard to pertinent characteristics of a house might be assumed to give a usable estimate of radon exposure, especially if living habits can also be taken into account. From the experiences so far, it seems as if the dominant sources of error for retrospective assessment of radon exposure have to do with poor knowledge of living habits, and also with the measurement and the determination of average air concentrations of radon (Bäverstam and Swedjemark, 1991).

So far, relatively little attention has been paid to the increased gamma radiation from those stone building materials that can also be a source of increased radon levels, even if the emanation from the ground usually predominates. As long as just radon and lung cancer are concerned, the gamma radiation aspect is probably not relevant, but it may be different for other diseases, such as acute leukaemia. There seems to be a need for disentangling the potential effects of indoor radon and gamma radiation, respectively, both with regard to leukaemia and other tumour types. It may be recalled in this context that there was a negative correlation between gamma radiation and various cancer types associated with radon exposure in the studies by Henshaw and co-workers (1990) as well as by others. On the other hand, there was a suggestion that background gamma radiation could be the cause of acute myeloid leukaemia in the studies by Flodin and co-workers (1981 and 1990).

For epidemiologic purposes it is important to find cheap and accurate methods to assess exposure over time. Comparisons with regard to some different methods for indoor measurements have been reported (Svensson et al., 1988). There remains a need to further evaluate the accuracy of the more or less crude estimations of past exposure that have to be applied for large numbers of subjects in epidemiologic studies. To this end, comparisons might perhaps be achievable with some detailed appraisals of the exposure for a few individuals. For a limited number of individuals it may also be possible to obtain some long-term reference estimates for indoor radon levels by studying alpha tracks on glass surfaces, which have been exposed for a long time in a home, e.g., mirrors or picture frame glass (Samuelsson, 1988).

Another aspect to consider would be how to cope with living habits, including the estimation of time spent indoors. The adjustments applied for occupancy in one study did not improve the dose-response relationships as could have been expected (Pershagen et al., 1992). In some of the studies on indoor radon, the study population has been restricted to rural areas in order to improve the precision in the assessment of individual exposure (Axelson et al., 1979; Edling et al., 1984). The rationale for this design was that the subjects in such a restricted study, especially the women, would be likely to have spent most indoor time in their homes and not in other constructions. Another aspect is that a rural population tends to be geographically very stable. There may be some areas or regions in the world where further studies along these lines might be undertaken.

An additional issue is the question whether the more recent or the more remote exposure might play the greater role for the risk of developing lung cancer in relation to radon exposure. In this respect a model based on miners, and provided in the BEIR IV report, emphasised the role of exposure 5–15 years before death from lung cancer. A reasonable agreement with this model could be seen for some data on indoor radon and lung cancer (Axelson et al., 1988; Axelson, 1991) but the agreement was less good with other data (Pershagen et al., 1992). There is an obvious need for further evaluations.

With regard to character of the disease, there is reason to consider the relationships for different histological types of lung cancer. Similarly, as for miners, the stronger relations to indoor radon exposure have appeared for the small cell and squamous types of lung cancer (Damber and Larsson, 1987; Pershagen et al., 1992).

From some of the studies referred to, it also seems as if cancers other than lung should be studied in relation to radon exposure. Particularly, the acute leukaemias, but also kidney cancers and possibly prostate cancer and melanoma, would be of interest in this respect. The development in molecular biology makes it possible to characterise lung cancers as well as various other neoplastic diseases with regard to genetic changes. As a consequence, more sensitive studies might be achieved, namely, if cancers with more or less specific changes could be more closely related to the exposure than other forms (Axelson and Söderkvist, 1991). A development along these lines has already been achieved when mutations, different from those usually associated with smoking, were found in the p53 gene in lung tumours among uranium miners (Vähäkangas et al., 1992).

Overall Effect of Exposure

Even if smoking today is accepted as the major cause of lung cancer, some critical comments from the past on the relation of smoking and lung cancer may be recalled (Passey, 1962; Burch, 1978; Burch, 1982). The remarks made could perhaps be taken to suggest an effect from some other widespread factor than smoking. Such an exposure could perhaps be indoor radon. Also the urban-rural gradient in lung cancer rates seems difficult to fully explain by differences in smoking. The influence of immigration on lung cancer morbidity is another matter that has not yet been fully understood. There is also a variation in lung cancer rates between countries, which sometimes remain when smoking is allowed (Dean, 1979). It may well be that differences in indoor radon exposure could be part of the explanation, and it might be worthwhile to reconsider the observations referred to when enough exposure data on indoor radon become available in the future.

An inverse relation of lung cancer and inhalation of cigarette smoke has been reported by some authors (Fisher, 1959; Higenbottam et al., 1982). Different explanations have been suggested for these findings, including a role for indoor radon (Wald et al., 1983; Axelson, 1984). Further studies would be necessary in order to confirm or refute the idea that indoor radon could help explain these rather odd observations regarding inhalation. In general, there is a need for more studies on the interaction of smoking and indoor radon exposure.

Subgroups at Special Risk

As yet, it is not clear whether specific subgroups should be given particular concern in terms of specific studies on the effect of indoor radon exposure. The interaction of radon and smoking certainly needs to be further evaluated, but it may also be speculated that occupational groups with exposure to various agents could be at increased risk if exposed to high levels of indoor radon in their homes. Information on occupation in any case would be desirable to control for confounding in studies of indoor radon exposure and lung cancer; also, the possibility of some interaction between indoor radon and occupational exposures might be evaluated.

Among uranium miners, a greater risk for lung cancer has been reported for shorter men (Archer et al., 1978). A relatively higher breathing rate in shorter men with smaller airways was assumed to be the consequence of the need for enough oxygen at hard work. As a result, the amount of inhaled radon progeny was thought to increase. Also, dosimetric models seem to suggest a greater risk to be associated with smaller airways, which might be of concern with regard to exposure during childhood and to some extent also with regard to women; the data are uncertain, however (BEIR IV, 1988).

The observations suggesting an effect of radon exposure with regard to childhood cancers (Henshaw et al., 1990) along with the dosimetric aspects certainly warrant specific studies in children. Also, prenatal exposure might be considered. There seems to be no information on how radon crosses the placenta, but it is assumed that the passage is free (BEIR IV, 1988). The sex ratio has been considered in two studies but the results were contradictory; no adverse effect has been noted with regard to reproductive outcomes in wives of uranium miners (IARC, 1988). Further studies seem to be warranted in these various respects.

HYPOTHESES FOR PRIORITY TESTING WITHIN THE RESEARCH NEEDS

In principle, the hypotheses for testing have to be set up by the individual researchers. Some guidelines might be given here, however, as based on this review and the needs discussed above. The following hypotheses, put forward in the form of broad questions and in arbitrary order regarding their priority, might be considered for testing in environmental epidemiology:

- Is the dose-response relationship for lung cancer essentially the same for miners and the general population?
- Are particular, for example, genetically characterised, tumour types associated with radon exposure?
- Is interaction between radon and smoking similar for the general population and miners?
- Do interactions occur between indoor radon exposure and various occupational exposures?
- Are cancer forms other than lung cancer associated with increased indoor radon exposure; does concomitant gamma exposure have no or only a marginal effect?
- Are children particularly sensitive to indoor radon exposure and can several tumour types be associated with this exposure?
- Do effects occur from prenatal exposure to indoor radon?
- Do relationships exist between indoor radon exposure and some biological marker, for example, 8-OHdG in urine?

REFERENCES

Åkerblom, G. and Wilson, C., Radon — geological aspects of an environmental problem. Rapporter och meddelanden nr 30. Sveriges geologiska undersökning, Uppsala, Sweden, 1982.

Albering, H. J., Hageman, G. J., Kleinjans, J. C. S., Engelen, J. J. M., Koulishcer, L., and Herens, C., Indoor radon exposure and cytogenetic damage, *Lancet,* 340, 739, 1992.

Alexander, F. E., McKinney, P. A., and Cartwright, R. A., Radon and leukaemia, *Lancet,* 335, 1136, 1990.

Archer, V. E., Association of lung cancer mortality with precambrian granite, *Occup. Environ. Health,* 42, 87, 1987.

Archer, V. E., Gillam, J. D., Lynn, M. A., and James, M. D., Radiation, smoking and height relationships to lung cancer in uranium miners, in *Proc. Third Int. Symp. Detection and Prevention of Cancer*, Nieburgs, H. E., Ed., Marcel Dekker, New York, 1978, 1689.

Arnstein, A., Über den sogenannten "Schneeberger Lungenkrebs", *Verh. dtsch. Ges. Pathol.,* 16, 332, 1913.

Axelson, O., Room for a role for radon in lung cancer causation?, *Med. Hypotheses,* 13, 51, 1984.

Axelson, O. and Sundell, L., Mining, lung cancer and smoking, *Scand. J. Work Environ. Health,* 4, 46, 1978.

Axelson, O., Edling, C., and Kling, H., Lung cancer and residency — A case referent study on the possible impact of exposure to radon and its daughters in dwellings, *Scand. J. Work Environ. Health,* 5, 10, 1979.

Axelson, O., Occupational and environmental exposure to radon: cancer risks, *Annu. Rev. Public Health,* 12, 235, 1991.

Axelson, O., Andersson, K., Desai, G., Fagerlund, I., Jansson, B., Karlsson, C., and Wingren, G., A case-referent study on lung cancer, indoor radon and active and passive smoking, *Scand. J. Work Environ. Health,* 14, 286, 1988.

Axelson, O. and Söderkvist, P., Characteristics of disease and some exposure considerations, *Appl. Occup. Environ. Hyg.,* 6, 428, 1991.

Battista, G., Belli, S., Garboncini, F., Comba, P., Giovanni, L., Sartorelli, P., Strambi, F., Valentini, F., and Axelson, O., Mortality among pyrite miners with low-level exposure to radon daughters, *Scand. J. Environ. Health,* 14, 280, 1988.

Bäverstam, U. and Swedjemark, G.-A., Where are the errors when we estimate radon exposure in retrospect?, *Rad. Protect. Dosimetry,* 36, 107, 1991.

Bean, J. A., Isacson, P., Hahne, R. M. A., and Kohler, J., Drinking water and cancer incidence in Iowa, *Am. J. Epidemiol.,* 116, 924, 1982.

BEIR IV. Committee on the Biological Effects of Ionizing Radiations, U.S. National Research Council, *Health Risk of Radon and Other Internally Deposited Alpha-Emitters,* National Academy Press, Washington, D.C., 1988.

Blot, W. J., Xu, Z.-Y., Boice Jr, J. D., Zhao, D.-Z., Stone, B. J, Sun, J., Jing, L.-B., and Fraumeni, Jr., J. F., Indoor radon and lung cancer in China, *J. Natl. Cancer Inst.,* 82, 1025, 1990.

Bridges, B. A., Cole, J., Arlett, C. F., Green, M. H. L., Waugh, A. P. W., Beare, D., Henshaw, D.L., and Last, R. D., Possible association between mutant frequency in peripheral lymphocytes and domestic radon concentrations, *Lancet,* 337, 1187, 1991.

Burch, P. R. J., Cigarette smoking and lung cancer, *Med. Hypotheses,* 9, 293, 1982.

Burch, P. R. J., Smoking and lung cancer. The problem of inferring cause. *J. R. Stat. Soc.,* 141, 4, 437, 1978.

Butland, B. K., Muirhead, C. R., and Draper, G. J., Radon and leukaemia, *Lancet,* 335, 138, 1990.

Castren, O., Mäkeläinen, I., Winqvist, K., and Voutilainen, A., Indoor radon measurements in Finland: A status report, in *Radon and Its Decay Products — Occurrence, Properties and Health Effects,* Hopke, P. K., Ed., American Chemical Society, Washington, D.C., 1987, 97.

Damber, L. A. and Larsson, L. G., Lung cancer in males and type of dwelling. An epidemiological pilot study, *Acta Oncol.,* 26, 211, 1987.

Damber, L. and Larsson, L. G., Combined effects of mining and smoking in the causation of lung cancer. A case-control study, *Acta Radiol. Oncol.,* 21, 305, 1982.

Dean, G., The effects of air pollution and smoking on health in England, South Africa and Ireland, *J. Ir. Med. Assoc.,* 72, 284, 1979.

Dickson, D., Home insulation may increase radiation hazard, *Nature, (London),* 276, 431, 1978.

Dousset, M. and Jammet, H., Comparison de la mortalitie par cancer dans le Limousin et le Poitou-Charentes, *Radioprotection GEDIM,* 20, 61, 1985.

Eatough, J. P. and Henshaw, D. L., Radon and prostate cancer, *Lancet,* 335, 1292, 1990.

Edling, C., Comba, P., Axelson, O., and Flodin, U., Effects of low-dose radiation — A correlation study, *Scand. J. Work Environ. Health,* 8, Suppl. 1, 59, 1982.

Edling, C., Kling, H., and Axelson, O., Radon in homes — A possible cause of lung cancer, *Scand. J. Work Environ. Health,* 10, 25, 1984.

Edling, C. and Axelson, O., Quantitative aspects of radon daughter exposure and lung cancer in underground miners, *Br. J. Ind. Med.,* 40, 182, 1983.

Fisher, R. A., *Smoking. The Cancer Controversy: Some Attempts to Assess the Evidence,* Oliver and Boyd, Edinburgh, 1959.

Fleischer, R. L., A possible association between lung cancer and phosphate mining and processing. *Health Phys.,* 41, 171, 1981.

Flodin, U., Andersson, L., Anjou, C.-G., Palm, U.-B., Vikrot, O., and Axelson, O., A case-referent study on acute myeloid leukaemia, background radiation and exposure to solvents and other agents. *Scand. J. Work Environ. Health,* 7, 169, 1981.

Flodin, U., Fredriksson, M., Persson, B., and Axelson, O., Acute myeloid leukaemia and background radiation in a expanded case-referent study. *Arch. Environ. Health,* 45, 364, 1990.

Forastiere, F., Quiercia, A., Cavariani, F., Miceli, M., Perucci, C. A., and Axelson, O., Cancer risk and radon exposure, *Lancet,* 339, 1115, 1992.

Forastiere, F., Valesini, S., Arca, M., Magliola, M. E., Michelozzi, P., and Tasco, C., Lung cancer and natural radiation in an Italian province, *Sci. Total Environ.,* 45, 519, 1985.

Fox, A. J., Goldplatt, P., and Kinlen, L. J., A study of the mortality of Cornish tin miners. *Br. J. Ind. Med.* 38, 378, 1981.

Härting, F. H. and Hesse, W., Der Lungenkrebs, die Bergkrankheit in den Schneeberger Gruben, *Vierteljahresschr. Gerichtl. Med. Oeff. Gesundheitswes.,* 30, 296, 1879.

Henshaw, D. L., Eatough, J. P., and Richardson, R. B., Radon as a causative factor of myeloid leukaemia and other cancers, *Lancet,* 335, 1008, 1990.

Hess, C. T., Weiffenbach, C. V., and Norton, S. A., Environmental radon and cancer correlation in Maine, *Health Phys.,* 45, 339, 1983.

Higenbottam, T., Shipley, M. J., and Rose, G., Cigarettes, lung cancer, and coronary heart disease: the effects of inhalation and tar yield, *J. Epidemiol. Comm. Health,* 36, 113, 1982.

Hodgson, J. T. and Jones, R. D., Mortality of a cohort of tin miners 1941-86, *Br. J. Ind. Med.,* 47, 665, 1990.

Hofmann, W., Katz, R., and Zhang, C., Lung cancer incidence in a Chinese high background area — epidemiological results and theoretical interpretation, *Sci. Total Environ.,* 45, 527, 1985.

Holaday, D. A., Digest of the proceedings of the 7-state conference on health hazards in uranium mining, *Arch. Ind. Health,* 12, 465, 1955.

Howe, G. R., Nair, R. C., Newcombe, H. B., Miller, A. B., Frost, S. E., and Abbatt, J. D., Lung cancer mortality (1950-1980) in relation to radon daughter in a cohort of workers at the Eldorado Beaverlodge uranium mine, *J. Natl. Cancer Inst.,* 77, 357, 1986.

Hultqvist, B., Studies on naturally occurring ionizing radiations with special reference to radiation dose in Swedish houses of various types, *Kungl svenska vetenskapsakademiens handlingar, 4:e serien, band 6, Nr 3,* Almqvist och Wiksell, Stockholm, Sweden, 1956.

IARC Monographs on the Evaluation of Carcinogenic Risks to Humans. Radon and Man-made Mineral Fibres, International Agency for Research on Cancer, Lyon, Frame, 1988.

ICRP, International Commission on Radiological Protection. Radiation Protection in Uranium and Other Mines. ICRP Publ. No. 24. Pergamon Press, Oxford, 1976.

Jörgensen, H. S., Lung cancer among underground workers in the iron ore mine of Kiruna based on thirty years of observation, *Ann. Acad. Med. (Singapore),* 13, 371, 1984.

Kinlen, L. J. and Willows, A. N., Decline in the lung cancer hazard: a prospective study of the mortality of iron ore miners in Cumbria, *Br. J. Ind. Med.,* 45, 219, 1988.

Klotz, J. B., Petix, J. R., and Zagraniski, R. T., Mortality of a residential cohort exposed to radon from industrially contaminated soil, *Am. J. Epidemiol.,* 129, 1179, 1989.

Kunz, E., Sevc, J., and Placek, V., Lung cancer in uranium miners, *Health Phys.,* 35, 579, 1978.

Kusnetz, H. L., Radon daughters in mine atmospheres — a field method for determining concentrations, *Ind. Hyg. Q.,* 17, 85, 1956.

Lanes, S. F., Talbott, E., and Radford, E., Lung cancer and environmental radon, *Am. J. Epidemiol.,* 116, 565, 1982.

Lawler, A. B., Mandel, J. S., Schuman, L. M., and Lubin, J. H., A retrospective cohort mortality study of iron ore (hematite) miners in Minnesota, *J. Occup. Med.,* 27, 507, 1985.

Lees, R. E. M., Steele, R., and Robert, J. H., A case-control study of lung cancer relative to domestic radon exposure, *Int. J. Epidemiol.,* 16, 7, 1987.

Letourneau, E. G., Mao, Y., McGregor, R. G., Semenciw, R., Smith, M. H., and Wigle, D. T., Lung cancer mortality and indoor radon concentrations in 18 Canadian cities, in *Proc. Sixteenth Midyear Topical Meeting of the Health Physics Society on Epidemiology Applied to Health Physics, Albuquerque,* Health Physics Society, Ottawa, 1983, 470.

Lubin, J. H., Qiao, Y., Taylor, P. R., Yao S.-X., Schatzkin, A., Mao, B.-L., Rao, J.-Y., Xuan, X.-Z., and Li, J.-Y., Quantitative evaluation of the radon and lung cancer association in a case control study of Chinese tin miners, *Cancer Res.,* 50, 174, 1990.

Lucie, N. P., Radon exposure and leukaemia, *Lancet,* 2, 99, 1989.

Lundin, Jr., F. E., Wagoner, J. K., and Archer, V. E., Radon daughter exposure and respiratory cancer. Quantitative and temporal aspects. *NIOSH-NIEHS Joint Monograph No. 1,* Public Health Service, Springfield, VA, 1971.

McGregor, R. G., Vasudev, P., Létourneau, E. G., McCullough, R. S., Prantl, F. A., and Taniguchi, H., Background concentrations of radon and radon daughters in Canadian homes, *Health Phys.,* 39, 285, 1980.

Mole, R. H., Radon and leukaemia, *Lancet,* 335, 1336, 1990.

Morrison, H. I., Semenciw, R. M., Mao, Y., Corkill, D. A., Dory, A. B., deVilliers, A. J., Stocker, H., and Wigle, D. T., Lung cancer mortality and radiation exposure among the Newfoundland fluorspar miners, in *Occupational Radiation Safety in Mining. Proc. Int. Conf.,* Stocker, H., Ed., Canadian Nuclear Association, Toronto, 1985, 354.

Muller, J., Wheeler, W. C., Gentleman, J. F., Suranvi, G., and Kusiak, R., Study of mortality of Ontario miners, in *Occupational Radiation Safety in Mining. Proc. Int. Conference,* Stocker, H., Ed., Canadian Nuclear Association, Toronto, 1985, 335.

Nero, A. V., Schwehr, M. B., Nazaroff, W. W., and Revzan, K. L., Distribution of airborne radon-222 concentrations in U.S. homes, *Science*, 234, 992, 1986.

Passey, R. D., Some problems of lung cancer, *Lancet*, 2, 107, 1962.

Pershagen, G., Liang, Z.-H., Hrubec, Z., Svensson, C., and Boice, J. D., Residential radon exposure and lung cancer in Swedish women, *Health Phys.*, 63, 179, 1992.

Pershagen, G., Damber, L., and Falk, R., Exposure to radon in dwellings and lung cancer: A pilot study, in *Indoor Air Radon, Passive Smoking, Particulates and Housing Epidemiology*, Berglund, B., Lindvall, T., and Sundell, J., Eds., Swedish Council for Building Research, Stockholm, 1984, 2, 73.

Placek, V., Smid, A., Sevc, J., Tomasek, L., and Vernerova, P., Late effects at high and very low exposure levels of the radon daughters, in Radiation research — somatic and genetic effects, in *Proc. 70th Int. Congress of Radiation Research*, Broerse, J. J., Barendsen, G. W., Kal, H. B., and Vanderkogel, A. J., Eds., Martinus Nijhoff, Amsterdam, 1983.

Qiao, Y., Taylor, P. R., Yao, S.-X., Schatzkin, A., Mao, B.-L., Lubin, J., Rao, J.-Y., McAdams, M., Xuan, X.-Z., and Li, J.-Y., Relation of radon exposure and tobacco use to lung cancer among tin miners in Yunnan province, China, *Am. J. Ind. Med.*, 16, 511, 1989.

Radford, E. P. and St. Clair Renard, K. G., Lung cancer in Swedish iron miners exposed to low doses of radon daughters, *N. Engl. J. Med.*, 310, 1485, 1984.

Roscoe, R. J., Steenland, K., Halperin, W. E., Beaumont, J. J., and Waxweiler, R. J., Lung cancer mortality among nonsmoking uranium miners exposed to radon daughters, *J. Am. Med. Assoc*, 262, 629, 1989.

Samet, J. M., Pathak, D. R., Morgan, M. V., Marbury, M. C., Key, C. R., and Valdivia, A. A., Radon progeny exposure and lung cancer risk in New Mexico U miners, *Health Phys.*, 56, 415, 1989.

Samet, J. M., Kutvirt, D. M., Waxweiler, R. J., and Key, C. R., Uranium mining and lung cancer in Navajo men, *N. Engl. J. Med.*, 310, 1481, 1984.

Samuelsson, C., Retrospective determination of radon in houses, *Nature (London)*, 334, 338, 1988.

Schmier, H. and Wick, A., Results from a survey of indoor radon exposures in the Federal Republic of Germany, *Sci. Total Environ.*, 45, 307, 1985.

Schoenberg, J. B., Klotz, J. B., Wilcox, G. P. N., Gil-del-Real, M. T., Stemhagen, A., and Mason, T. J., Case-control study of residential radon and lung cancer among New Jersey women, *Cancer Res.*, 50, 6520, 1990.

Simpson, S. G. and Comstock, G. W., Lung cancer and housing characteristics, *Arch. Environ. Health*, 38, 248, 1983.

Stranden, E., Radon-222 in Norwegian dwellings, in *Radon and Its Decay Products-Occurrence, Properties and Health Effects*, Hopke, P. K., Ed., American Chemical Society, Washington, D.C., 1987, 70.

Stranden, E., Berteig, L., and Ugletveit, F., A study on radon in dwellings, *Health Phys.*, 36, 413, 1979.

Svensson, C., Pershagen, G., and Hrubec, Z., A comparative study on different methods of measuring Rn concentrations in homes, *Health Phys.*, 55, 895, 1988.

Svensson, C., Pershagen, G., and Klominek, J., Lung cancer in women and type of dwelling in relation to radon exposure, *Cancer Res.*, 49, 1861, 1989.

Svensson, C., Eklund, G., and Pershagen, G., Indoor exposure to radon from the ground and bronchial cancer in women, *Int. Arch. Occup. Environ. Health*, 59, 123, 1987.

Swedjemark, G. A., Buren, A., and Mjönes, L., Radon levels in Swedish homes: A comparison of the 1980s with the 1950s, in *Radon and Its Decay Products — Occurrence, Properties and Health Effects,* Hopke, P. K., Ed., American Chemical Society, Washington, D.C., 1987, 85.

Tagesson, C., Källberg, M., and Leanderson, P., Determination of urinary 8-hydroxydeoxyguanosine by coupled-column high-performance liquid chromatography with electrochemical detection: a noninvasive assay for in vivo oxidative DNA damage in humans, *Toxicol. Methods,* 1, 242, 1992.

Tomasek, L., Darby, S., Swerdlow, A. J., Placek, V., and Kunz, E., Radon exposure and cancers other than lung cancer among uranium miners in West Bohemia, *Lancet,* 341, 919, 1993.

Vähäkangas, K. H., Samet, J. M., Metcalf, R. A., Welsh, J. A., Bennett, W. P., Lane, D. P., and Harris, C. C., Mutations of p53 and ras genes in radon-associated lung cancer from uranium miners, *Lancet,* 339, 576, 1992.

Wagoner, J. K., Miller, R. W., Lundin, Jr., F. E., Fraumeni, J. F., and Haij, N. E., Unusual mortality among a group of underground metal miners, *N. Engl. J. Med.,* 1963, 269, 281, 1963.

Wald, N. J., Idle, M., Boreham, J., and Bailey, A., Inhaling and lung cancer: an anomaly explained, *Br. Med. J.,* 287, 1273, 1983.

Waxweiler, R. J., Wagoner, J. K., and Archer, V. E., Mortality of potash workers, *J. Occup. Med.,* 15, 486, 1973.

Waxweiler, R. J., Roscoe, R. J., Archer, V. E., Thun, M. J., Wagoner, J. K., and Lundin, E., Mortality follow-up through 1977 of the white underground uranium miners cohort examined by the U.S. Public Health Service, in *Radiation Hazards in Mining: Control, Measurement and Medical Aspects,* Gomez, M., Ed., Society of Mining Engineers of the American Institute of Mining, Metallurgical, and Petroleum Engineers, New York, 1981, 823.

Wilkinson, G. S., Gastric cancer in New Mexico counties with significant deposits of uranium, *Arch. Environ. Health,* 40, 307, 1985.

3

ASBESTOS

Benedetto Terracini

CONTENTS

INTRODUCTION

Asbestos (CAS 1332-21-4) is the generic name given to a class of natural fibrous silicates that vary considerably in their physical and chemical properties. The four major forms of asbestos are chrysotile or white asbestos (12001-29-5), amosite (12172-73-5), anthophyllite (17068-78-9), and crocidolite or blue asbestos (12001-28-4). Less commercially important forms are actinolite (13768-00-8) and tremolite (14567-73-8).

Data on properties and industrial uses of asbestos from different sources are well summarised.[1] Chrysotile, amosite, and particularly crocidolite have extremely high-tensile strengths and are used extensively as reinforcers in cements, resins, and plastics. Chrysotile fibers are soft and flexible, whereas crocidolite and amosite fibers are hard and brittle. Although chrysotile is most adaptable to industrial use, the combination of chrysotile with crocidolite and amosite is particularly useful for adding specific properties such as rigidity. Because of its flexibility and softness, chrysotile is particularly suitable for being spun into textiles.

Asbestos is or has been used in more than 5000 products. Longer, maximum-strength fibers are used in the production of textiles, electrical insulation, and pharmaceutical and beverage filters. Medium-length fibers are used in the production of asbestos-cement pipe and sheet, clutch facings, brake linings, asbestos paper, packaging, gaskets, and pipe coverings. Short fibers are used as a reinforcer in plastics, floor tiles, coatings and compounds, papers, and roofing felts. Uses in the U.S. in 1987 were 22% in friction products, 17% in roofing products, 17% in asbestos cement, 15% in coating and compounds, 11% in paper, and 18% in other uses.[1]

During the last decade or so, at least in developed countries, consumption of asbestos has declined. In the U.S.,[1] domestic production was 176 million lb in 1980 whereas in 1989 two firms produced (in terms of sales) 37.4 million lb. Corresponding figures for imports were 721 million lb in 1980 vs. 121 million lb in 1989; exports did not vary as greatly as imports (108 and 50 million lb, respectively, in 1980 and 1989).

EXPOSURE

Worker exposure is a concern in the mining and milling of asbestos, during the manufacture of all asbestos products, and in the construction and ship-building industries. However, exposure occurs also in occupations involving use of asbestos end products, such as insulation, brake repair and maintenance, building demolition, and others. Attempts of overall occupational risk assessment are impaired by limited knowledge on nature of products prepared with asbestos and their industrial uses. A survey in the U.S. estimated that in the early eighties over 150,000 workers (including 7603 women) were potentially exposed to asbestos in the workplace.[2]

The 1986 threshold limit values as 8-hr time-weighted averages (TWAs) established by the American Conference of Governmental Industrial Hygienists (ACGIH) for amosite, chrysotile, crocidolite, and other forms of asbestos were (in fibers per cubic centimeter), respectively, 0.5, 2, 0.2, and 2. Currently, the limit value in Italy is 0.2 fibers per cubic centimeter, with the exception of chrysotile (in the absence of other forms of asbestos), for which the limit is 0.6 (with the exception of extracting activities, for which the previous limit of 1.0 will operate until 1996.[3]) At least in industrial countries, the intensity of occupational exposure has decreased in the last 30–50 years, but levels in the general environment (air, water, food, etc.) have increased. Asbestos is used so widely that the entire population is potentially exposed to some degree. Application of asbestos materials to buildings (and building demolition) as well as vehicle brake linings and wearing of asbestos-containing materials accounts for a significant amount of emissions to the environment. In the early seventies, atmospheric concentrations up to 100 ng/m^3 (usually not exceeding 10 ng/m^3) were reported in the general urban atmosphere[4] vs. ambient air concentration < 0.01 ng/m^3 in country areas, away from anthropogenic or natural sources of asbestos.[5]

The general population living in the vicinity of mines and processing plants may be unduly exposed to asbestos through a variety of sources and not only directly from emissions into the atmosphere. Disposal of mining and industrial (and building material) wastes contributes to asbestos exposure as well as the domestic exposure in houses of workmen, where concentrations on the order of 100–500 ng/m^3 have been reported.[5]

Legislation providing for cessation of extraction and industrial uses of asbestos before the end of the nineties has passed in some countries, including Italy.[6]

EVIDENCE OF CARCINOGENICITY

Experimental Systems

There is sufficient evidence for the carcinogenicity of the chrysotile, amosite, anthophyllite, and crocidolite in experimental animals.[1,4,7-10] When administered by inhalation, they all produced mesotheliomas and lung carcinomas in rats, whereas intrapleural administration produced mesotheliomas in rats and hamsters. Intraperitoneal administration of chrysotile, crocidolite, and amosite induced peritoneal tumours in mice and rats. When filter material containing chrysotile was added to the diet, a statistically significant increase in the incidence of malignant tumours was observed in rats. However, tumour incidence was not increased by oral administration of amosite in rats and hamsters or of chrysotile in hamsters. In rats of either sex, dietary addition of short- or intermediate-range fiber length pelleted chrysotile was not carcinogenic; however, males receiving intermediate-range fibers showed a moderate increase of benign polyps of the large intestine. Cocarcinogenesis studies of 1,2-dimethylhydrazide dihydrochloride and intermediate-range chrysotile led to inconclusive results.

Epidemiological Evidence

There is sufficient evidence for the carcinogenicity of asbestos, chrysotile, amosite, anthophyllite, and crocidolite.[1,10] Occupational exposure to chrysotile, amosite, anthophyllite, and mixtures containing crocidolite has resulted in a high incidence of lung cancer. Mesotheliomas have been observed after occupational exposure to crocidolite, amosite, and chrysotile asbestos: the risk may be lower among workers exclusively exposed to chrysotile.[11] Mesotheliomas have occurred in individuals living in the neighborhood of asbestos factories and mines and in people living with asbestos workers,[12-17] but risks could be estimated only from some of these studies, and were based on small absolute numbers of cases. An excess of laryngeal cancer in occupational groups has been consistently reported whereas reports of excesses of gastrointestinal cancer have been inconsistent. No clear excess of cancer has been associated with the presence of asbestos fibers in drinking water. It is a very well-known fact that both cigarette smoking and occupational exposure to asbestos fibers increase lung cancer incidence independently and that they act multiplicatively, if present together.[18-19] Smoking does not seem to modify the ability of asbestos to induce mesothelioma.[18]

OTHER NOXIOUS EFFECTS OF ASBESTOS

The presence of asbestos fibers in the lungs may result in a type of incapacitating fibrosis known as asbestosis. At autopsy, pulmonary morphologic changes in asbestosis are characteristic. Clinical and radiological findings may be indistinguishable from other forms of pulmonary fibrosis. Mild pulmonary fibrosis following limited exposure to asbestos may be asymptomatic, unless lung function is also affected by other causes (such as tobacco smoke or chronic obstructive disease). Dose-response relationships in an asbestos textile factory led to the estimate that an annual incidence of 0.5% (SE 0.2) certified asbestosis has occurred after cumulative doses <100 fiber years per ml, i.e., within a working life at 2 fibers per ml. The annual rates of incidence for possible asbestosis and crepitations are higher at any given level and suggest the possible occurrence of asbestos-related disease at cumulative doses of less than 50 fiber years per ml.[20,21]

It has long been known[22] that autopsy or radiological findings of pleural plaques are associated with (but not exclusively caused by) asbestos. They may be useful indicators of previous exposure but are of no clinical importance.

RESEARCH NEEDS

Role of Vital Statistics or Available Disease-Oriented Databases in Hypothesis Generating or Testing

Adequate records of the occurrence of mesotheliomas may provide an important lead to the identification of subpopulations exposed (at least in the past) to asbestos. The only other etiological agent known for this cancer is ionising radiation.[23] In industrialised countries, a conspicuous proportion of mesotheliomas are attributable to occupational exposure to asbestos (well over 50% in areas characterised by industrial activities entailing extraction/use/exposure to asbestos). As mentioned above, in the same areas, mesotheliomas can be caused also by environmental (nonoccupational) exposure to asbestos.

Descriptive analyses based on vital statistics are of much more limited use for generating or testing hypotheses related to other asbestos-related cancers, such as lung cancer. This simply reflects the limited potential of such analyses for controlling confounders such as tobacco smoke and diet.

Incidence data on mesotheliomas can be collected through cancer registries (where in operation) or *ad hoc* surveys, of which some good examples have been reported.[16,24] Legislation of the European Community[25] (see article 17) provides for the implementation of national registries of cases of mesothelioma and asbestosis. In Italy, words have been changed into "ascertained cases of asbestos-related mesothelioma and asbestosis."[3] In fact, the process is intended to include both recording of cases and discrimination between asbestos "related" and "unrelated" cases (see "Exposure Assessment" below.) Criteria for optimal and exhaustive population-based recording of cancer cases are available.[26] Potentially, they are applicable to registration of one single (albeit rare) cancer, such as mesothelioma, also in areas with no experience of cancer registration, but preliminary feasibility investigations are required.

The provision for registration of cases of asbestosis involves the assessment of exhaustiveness to which this occupational disease is currently recognised and compensated. In addition, there is a need for estimating reproducibility of diagnosis in the case of mild asbestosis.

In some European countries, descriptive epidemiology of mesothelioma based on mortality statistics provided useful information on geographic and temporal differences.[27,28] Since the availability of mortality statistics is widespread and dates back several decades, analyses of the occurrence of mesothelioma can be performed rapidly and cheaply. They deserve high priority, particularly if complemented by estimates of the reliability of certification of causes of death.

Endpoints of population-based descriptive analyses on mesothelioma occurrence include: (1) cluster identification and broad quantification of the excess (ideally, the cluster, if confirmed, should be sufficiently circumscribed as to warrant additional studies aimed at investigating causality of its association to a hypothesised source); (2) assessment of time trends, either upwards, as a reflection of past exposure, or downwards, as a measure of effectiveness of environmental control; (3) provision of population databases for epidemiological analytical studies; and (4) development of know-how for geographic analyses of disease distribution.

Role of Other Databases in Hypothesis Generating or Testing

A number of exhaustive occupational cohorts of asbestos workers are available throughout the world. The majority were assembled retrospectively, and information on previous concentration of asbestos in the workplace was not always adequate. For cohorts engaged in the manufacture of asbestos products, the information on the types of asbestos also was often unsatisfactory. Finally, individual exposure to confounders (other occupational carcinogens, tobacco smoking, and dietary habits in the first place) was often incomplete. Nevertheless, the available cohorts have led not only to the identification of causality (g), but also to hypotheses on mechanisms of action and to mathematical models for lung and pleural carcinogenesis.[29]

For most cohorts, epidemiologists taking care of the follow-up have already considered means for integrating demographic data with relevant information on individuals' exposure to asbestos and to confounders. However, it remains to be checked whether for all major cohorts all means have been assayed for making the best use of records kept by firms, formal inquiries with the workforce, or informal contacts *in loco*. An accurate reconstruction and recording of the technologies (either extraction or production) used in different periods might lead to more refined (or, perhaps, fresh) estimates of concentration and nature of asbestos present in the workplace.

Such historical assessment deserves high priority: the cessation of asbestos-related industrial activities in Western Europe might bring about a loss of valuable data. In industrialised countries, the diffusion of activities entailing occupational exposure to asbestos and the relative high proportion of exposed persons renders the case-control approach particularly suitable for etiological investigations. In fact, large disease-oriented databases complemented by systematic collection of occupational histories are already available. The need is for systematic criteria for assessing individual exposures (see next section).

Exposure Assessment

Workplace

In a number of plants engaged in asbestos-related activities (extraction, production of brakes and other friction materials, asbestos cement, shipyards, etc.), environmental analyses have been routinely carried out in the last decade or two. Comparing findings in different industrial settings would be useful for estimating the error of indirect estimates in workplaces where (or in periods in which) no analyses have been carried out.

Estimating individual exposures in more remote periods or in other (less conventional) activities entailing exposure to asbestos — if and when relevant — is problematic. Probably, the best approach is that based on subjective probability, through assignment of scores by a number of "experts", complemented by assessment of inter- and intra-observer reproducibility and comparison with workplaces where measurements are available.[30] The exercise is promising but requires methodological refinement.

General Environment

Given the variety of (and limited knowledge on) work-related uses of asbestos, criteria are needed in order to assess the reliability of judgement that a person (or a group) *has not* experienced occupational exposure to asbestos. Such a judgement, usually based on exhaustive recall of the individual's occupational history, requires thorough knowledge of industrial technology in a given area. The problem is specular to that mentioned in the previous paragraph; again, it is probably best approached through subjective probability, assignment of scores, and measurement of reproducibility within a number of experts.

Environmental measurements of asbestos in highly polluted areas are important for risk estimates and for interpreting case reports of mesotheliomas in persons not exposed in the workplace (other asbestos-related conditions are less relevant). Sampling and analytical methods are fairly standardised (although inter-laboratory reproducibility of results has been investigated to a limited extent). In recent times, the logistics of their implementation in "suspected" areas has posed no major problems.

The problem is much more complex in areas suspected of having experienced asbestos pollution in the general environment in periods for which no measurements are available. Still, these circumstances may help in understanding dose-response ("high" nonoccupational exposures in the general environment are likely to be lower than occupational exposures in many settings) and other biological aspects of asbestos toxicity, such as the induction of mesotheliomas following exposure during childhood, which does not seem to fit into the models developed in studies in the occupational environment.[31]

Thus, criteria for estimating previous environmental asbestos pollution in the general environment must be developed and experiences in different areas should be compared. The exercise might be expanded to the quantification of fiber levels and asbestos bodies in the lungs of residents in polluted areas,[32] due to attention being given to standardisation of the methods and evaluation of bias inherent to the selection of lung specimens which may be available.

Further areas worth being explored are (1) the occurrence of indicators of exposure to asbestos in domestic animals and (2) prevalence of pleural plaques in routine chest X-rays from asymptomatic persons.

Mechanism of Action

Mathematical models for lung and pleural carcinogenesis have been developed and verified in some databases.[29] Inconsistencies under other circumstances do not necessarily indicate inadequacy of the models insomuch as they might reflect inadequate estimates of intensity of exposure, different effects of the various forms of asbestos, or insufficient control of modifying effects of tobacco. There is a need to verify the applicability of the models in occupational cohorts for which exposure is relatively well characterised, as well as in environmentally exposed populations.

Definition of Disease Study Entities, Including Aspects Regarding Classification

In studies on the association between asbestos exposure and cancer, it has long been known[18,33] that death certificates underestimate the occurrence of mesothelioma and — to a certain extent — lung cancer. The bias can be partly overcome through the introduction of the criterion of the "best evidence" (from clinical, pathological, or post-mortem records, as applicable to each individual case). Nevertheless, this creates a further bias in the comparison between observed and expected numbers of cases, since the latter cannot be adjusted for degree of evidence. In addition, the few available studies on inter-observer agreement in histological typing of lung cancer[34] suggest that agreement is far from being 100%, the highest disagreement being achieved for adenocarcinoma.

Comparability of level of diagnostic reliability between currently available databases has not been adequately assessed: the possibility of a multicentric coordination should be explored.

Subgroups at Special Risk

Smokers

The multiplicative effect of lung cancer risks from cigarette smoke and from exposure to asbestos is well established. There are no clear-cut data on two issues with major practical implications: (1) whether interaction occurs to the same extent for all types of asbestos and (2) whether the asbestos-related excess risk from cigarette smoke persists *after* cessation of asbestos exposure. In the perspective that extraction and uses of asbestos will be abandoned (at least in some countries), clarifying this issue would provide guidance for setting priorities in anti-smoking campaigns. It might also contribute to interpretation of respective roles of asbestos and tobacco in mesothelioma initiation and promotion.

Individual Susceptibility

Some series of mesotheliomas other than those occurring in asbestos-related occupations (but possibly associated to asbestos in the general environment) have included children or relatively young people.[35,36] No hypotheses can be forwarded on whether this reflects level of exposure, individual susceptibility (genetically or nongenetically determined), or chance. The family pedigree of these persons might be informative.

Human mesotheliomas are highly aneuploid and show a wide variety of chromosomal changes, but some nonrandom structural and numerical changes have been identified, including changes on chromosome 1: it has been postulated that this mechanism might play a role in the genesis of asbestos-related cancer.[37,38] Finally, an area as yet barely explored is oncogene expression in mesothelioma.

These aspects could be further explored through a closer cooperation between epidemiologists and basic science investigators.

REFERENCES

1. U.S. Department of Health and Human Services. Public Health Service: Sixth Annual Report on Carcinogens — Summary 1991. Prepared for the National Institute of Environmental Health Sciences, Research Triangle Park, NC by Technical Resources Inc. Rockville, MD (contract N01 ES 3 5025).
2. NIOSH — National Institute for Occupational Safety and Health. National Occupational Exposure Survey 1980-83. Department of Health and Human Services, Cincinnati, Ohio, 1984.
3. Decreto Legislativo 15 agosto 1991 n. 277 Attuazione delle direttive n. 80/1107/CEE, n. 82/605/CEE, n. 83/477/CEE, n. 86/188/CEE e n. 82/642/CEE in materia di protezione dei lavoratori contro i rischi derivanti da esposizione ad agenti chimici, fisici e biologici durante il lavoro, a norma dell'art. 7 della legge 30 luglio 1990 n. 212. Supplemento ordinario` n. 200, *Gazzetta Ufficiale della Repubblica Italiana,* 27 agosto 1991.
4. International Agency for Research on Cancer. *IARC Monographs on the Evaluation of the Carcinogenic Risk of Chemicals to Man. Vol. 14: Asbestos,* IARC, Lyon, France, 1977.
5. *Chem. Eng. News,* 63, n. 9, 1985.
6. Legge 27 marzo 1992 n. 257 Norme relative alla cessazione dell'impiego dell'amianto. Supplemento ordinario` n. 87, *Gazzetta Ufficiale della Repubblica Italiana,* 13 aprile 1992.
7 International Agency for Research on Cancer. *IARC Monographs on the Evaluation of the Carcinogenic Risk of Chemicals to Man. Vol. 2: Some Inorganic and Organometallic Compounds,* IARC, Lyon, France, 1973.

8. International Agency for Research on Cancer. *IARC Monographs on the Evaluation of the Carcinogenic Risk of Chemicals to Humans: Chemicals and Industrial Processes Associated with Cancer in Humans*. Supplement 1, IARC, Lyon, France, 1979.

9. International Agency for Research on Cancer. *IARC Monographs on the Evaluation of the Carcinogenic Risk of Chemicals to Humans: Chemicals and Industrial Processes and Industries associated with Cancer in Humans. Supplement 4*, IARC, Lyon, France, 1982.

10. International Agency for Research on Cancer. *IARC Monographs on the Evaluation of the Carcinogenic Risk of Chemicals to Humans: Overall Evaluations of Carcinogenicity. Supplement 7*, IARC, Lyon, France, 1987.

11. Hughes, J. M., Weill, H., and Hammad, Y. Y., Mortality of workers employed in two asbestos-cement manufacturing plants. *Br. J. Ind. Med.* 44, 161-174, 1987.

12. Wagner, J. C., Sleggs, C. A., and Marchand, P. Diffuse pleural mesothelioma and asbestos exposure in the North Western Cape Province. *Br. J. Ind. Med.* 17, 260-271, 1960.

13. Newhouse, M. A. and Thompson, H., Mesothelioma of pleura and peritoneum following exposure to asbestos in the London Area. *Br. J. Ind. Med.* 22, 261-269, 1965.

14. Vianna, N. J. and Polak, A. K., Non-occupational asbestos exposure and malignant mesotheliom in females. *Lancet* 1, 1061-1063, 1987.

15. Anderson, H., Family contact exposure. In *Proc. of the World Symposium on Asbestos* Montreal May 25-27 1982, Canadian Asbestos Information Centre. pp. 349-362, 1982.

16. Magnani, C., Borgo, G., Betta, G. P., Botta, M., Ivaldi, C., Mollo, F., Scelsi, M., and Terracini, B., Mesothelioma and non-occupational environmental exposure to asbestos (letter). *Lancet* 338, 949, 1991.

17. Magnani, C., Terracini, B., Ivaldi, C., Botta, M., Budel, P., Mancini, A., and Zanetti, R., A cohort study on mortality among wives of workers in the asbestos-cement industry in Casale Monferrato, Italy. *Br. J. Ind. Med.* 50, 779-784, 1993.

18. Hammond, E. C., Selikoff, I. J., and Seidman, H., Asbestos exposure, cigarette smoking and death rates. *Ann. N.Y. Acad. Sci.* 330, 473-490, 1979.

19. Saracci, R., The interactions of tobacco smoking and other agents in cancer etiology. *Epidemiol Rev.* 9, 175-193, 1987.

20. Acheson, E. D. and Gardner, M., The ill effectos of asbestos on health. Asbestos: Final Report to the Advisory Committee 2. HMSO, London, 1979.

21. Berry, G., Gilson, J. C., Holmes, S., Lewinsohn, H. C., and Roach, S. A., Asbestosis: a study of dose-response relationships in an asbestos textile factory. *Br. J. Ind. Med.* 36, 98-112, 1979.

22. Meurman, L., Asbestos bodies and pleural plaques in a Finnish series of autopsy cases. *Acta Pathol. Microbiol. Scand.* Suppl 181, 1-107, 1966.

23. Patterson, J. T., Greenberg, S. D., and Buffler, P. A., Non-asbestos-related malignant mesothelioma. A review. *Cancer* 54, 951-960, 1984.

24. Begin, R., Gauthier, J. J., Desmeules, M., and Ostiguy, G., Work-related mesothelioma in Quebec, 1967-1990. *Am. J. Industr. Med.* 22, 531-542, 1992.

25. Council of the European Communities, Directive 83/477, September 19, 1983 (Title in Italian is Sulla protezione dei lavoratori contro i rischi connessi con esposizione all'amianto durante il lavoro).
26. MacLennan, R., Muir, C., Steinitz, R., and Winkler, A. Eds. *Cancer and Its Techniques*. IARC Sci. Publ. n. 21, pp. 1-235, Lyon, France, 1978.
27. Gardner, M. J., Acheson, E. D., and Winter, P. D., Mortality from mesothelioma of the pleura during 1968-78 in England and Wales. *Br. J. Cancer* 46, 81-88, 1982.
28. Di Paola, M., Mastrantonio, M., Comba, P., Grignoli, M., Maiozzi, P., and Martuzzi, M., Distribuzione territoriale della mortalità per tumori maligni della pleura in Italia. *Ann. Ist. Sup. San.* 28, 589-600, 1992.
29. Doll, R. and Peto, J., Effects on health of exposure to asbestos. Health & Safety Commission, HMSO, pp. 1-58, London, 1985.
30. Siemiatycki, J. Ed. *Risk Factors for Cancer in the Workplace*. CRC Press, Boca Raton, FL, 1991.
31. Peto, J., Fibre carcinogenesis and environmental hazards. In *Non-occupational Exposure to Mineral Fibers*. Bignon, J., Peto, J., and Saracci, R. Eds. IARC Sci. Publ. n. 90, pp. 457-470, Lyon, France, 1989.
32. Case, B. W. and Sebastien, P., Fibre levels in lung and correlation with air samples. In *Non-occupational Exposure to Mineral Fibers*. Bignon, J., Peto, J., and Saracci, R. Eds. IARC Sci. Publ. n. 90, pp. 207-218, Lyon, France, 1989.
33. Newhouse, M. L. and Wagner, J. C., Validation of death certificates in asbestos workers. *Br. J. Ind. Med.* 26, 302-307, 1969.
34. Campobasso, O., Andrion, A., Ribotta, M., and Ronco, G., The value of the 1981 WHO histological classification in inter-observer reproducibility and changing pattern of lung cancer. *Int. J. Cancer* 53, 205-208, 1993.
35. Wassermann, M., Wassermann, D., Steinitz, R., Katz, L., and Lemesch, C., Mesothelioma in children. In *Biological Effects of Mineral Fibers*, Wagner, J. C. Ed. International Agency for Research on Cancer Sci. Publ. n. 30, pp. 253-257, Lyon, France, 1980.
36. Armstrong, B. K., Musk, A. W., Baker, J. E., Hunt, J. M., Newall, C. C., Henzell, H. R., Blundson, B. S., Clarke-Hundley, M. D., Woodward, S. D., and Hobbs, M. S. T., Epidemiology of malignant mesothelioma in Western Australia. *Med. J. Aust.* 141, 86-88, 1984.
37. Walker, C., Everitt, J., and Barrett, J. C., Possible cellular and molecular mechanisms for asbestos carcinogenicity. *Am. J. Ind. Med.* 21, 253-273, 1992.
38. Barrett, J. C., Mechanisms of action of known human carcinogens. In *Mechanisms of Carcinogenesis in Risk Identification*. Vainio, H., Magee, P. N., McGregor, D. B., and McMichael, A. J. Eds. International Agency for Research on Cancer Sci. Publ. n. 116, pp. 115-134, 1992.

4   EVALUATION OF RISKS ASSOCIATED WITH HAZARDOUS WASTE

Anthony B. Miller

CONTENTS

INTRODUCTION

At the request of the Agency for Toxic Substances and Disease Registries (ATSDR) and with support from the U.S. Environmental Protection Agency (EPA), the National Research Council (NRC) in the U.S. convened the Committee on Environmental Epidemiology to suggest how to improve the scientific bases for evaluating effects of environmental pollution on public health. Our first report reviewed the published literature on hazardous waste sites up to mid-1990 and considered the findings in relation to the legislative mandate of EPA and ATSDR (National Research Council, 1991). The second report of the committee will identify research opportunities and methodological issues for the general field of environmental epidemiology (National Research Council, 1995). In addition, we shall expand our previous literature review, including reports becoming accessible regarding environmental contamination in Eastern Europe, and reports from public health departments and court judgements that are not available in the general scientific literature that we were able to access, which we refer to as the "grey" literature.

Our charge was not to consider radiation-related issues, those arising both from ionising and nonionising radiation (including in the latter category possible adverse effects of electrical and magnetic fields), toxic air pollutants, or issues related to pesticide exposure in children. These important topics are being covered by other NRC committees.

ATSDR was set up in 1983 as a result of the legislation entitled the Comprehensive Environmental Response, Compensation and Liability Act, passed in 1980, that also established the Superfund to manage and ameliorate the effects of hazardous waste sites. Unfortunately, the application of this legislation in practice created a risk that public health is being imperiled. This risk derives in part from the fact that much of the information collected under the Superfund program was largely used to manage hazardous waste sites but not to evaluate human exposure, and thus to identify and ameliorate those exposures truly hazardous to human health (GAO, 1991). While billions of dollars have been spent on hydrogeological site characterisation, few funds have been applied to the development of an active public health program. Moreover, many relevant sources of human exposure from such sites have not been identified, so that no remedial action is possible for such exposures. There is also a critical need for funds to expand the research mission of ATSDR to more fully cover public health issues and exploit research opportunities.

THE DIMENSIONS OF THE PROBLEM

There is no question that substances toxic to humans and several animal species abound in hazardous waste sites. The human health studies reported in the scientific literature have shown that serious health effects cannot be ruled out (National Research Council, 1991). In the U.S., more than 6 billion tons of waste is produced annually — nearly 50,000 tons per person. One survey (ATSDR, 1989) estimated that more than 40 million people live within 4 miles and about 4 million within 1 mile of a Superfund site. Residential proximity itself, however, does not mean that exposures and health risks are occurring, only that the potential for exposure is increased. Further, the real dimensions of the problem are unknown. Decisions have been taken not to list some sites in the National Priority List (NPL) of the Superfund even though these sites have never been fully characterised as to the substances they contain or the potential for human exposure. More than 31,000 sites have been reported that potentially need cleanup, but it has been estimated that the real total may approximate to 440,000 (OTA, 1989). As of 1989, the EPA had completed more than 27,000 preliminary assessments and detailed investigations on more than 9000 sites, with 1189 on the NPL in 1991. However, the ATSDR found that data adequate for evaluating environmental contamination and public health risks were only available for 31% of 951 NPL sites assessed (ATSDR, 1989).

BACKGROUND OF RESEARCH IN ENVIRONMENTAL EPIDEMIOLOGY RELEVANT TO PUBLIC HEALTH

The causes of some increasingly common chronic diseases in developed countries remain largely unknown. Thus, more than 60% of all birth defects are of unknown etiology, as are many degenerative neurological diseases. While genetic factors are clearly involved in many of these diseases, the role of environmental factors needs to be clarified. Unfortunately, the nature of the patterns in time and place of noncommunicable, chronic diseases often cannot be determined, because appropriate systems have not been developed to evaluate such patterns. A top priority for environmental health research thus becomes the identification of environmental factors that contribute to diseases of unknown etiology that currently have or may in the future have an important effect on the quality of life. Various routes of exposure may be relevant to some of these diseases, including natural and human-made radiation, air, water, and ingestion.

Among endpoints that need to be assessed are asthma and adult onset respiratory hypersensitivity; lung function and growth; degenerative neurological diseases; neurobehavioral and developmental deficits; birth defects, including those for which male-mediated exposures might be important; reproductive health in males and females, including changes in development of menses, menopause, sexual functioning, and endometriosis; and immunologic and endocrine diseases, such as diabetes.

There is a mix of infectious and chronic diseases in developing and developed countries. While different types of diseases predominate, the range of diseases found in both is remarkably similar. Thus, for both infectious and chronic diseases, basic sanitation and living conditions remain important risk factors, and include urban infrastructure, proper housing, clean air and water, and clean and safe working conditions. All of these are threatened by the sanitary and toxicologic burden that is being accumulated in all countries (World Bank, 1992). These are the traditional areas where hygienists have had a great impact in the past in developed countries.

DETERMINING RATES OF OCCURRENCE OF DISEASES OF UNKNOWN ETIOLOGY

For many diseases of potential interest in environmental epidemiology, routinely available data do not include rates of occurrence in defined populations, or anything approximating disease incidence. This may make it impossible to determine whether incidence has changed in response to a new or changing environmental exposure. Although incidence rates may not be required under circumstances where a gradient of exposure can be determined and risk of disease evaluated in a dose-response mode in a specially designed study, for many purposes before an assessment can be made of the relationship of diseases of environmental etiology to environmental factors, baseline rates need to be determined.

A variety of tools are available for estimating baseline rates, including: the validation of biologic markers to identify early stages of pathology for specific disease processes; reliable biomonitoring of relevant populations believed to be at risk; routine evaluations of sources of data such as in the U.S., large-scale Health Maintenance Organisations, Health Care Financing Administration records, and health insurance information from which rates of treatment and incidence might be inferred for specific populations; improved use of registries of specific diseases, such as those being developed by the ATSDR; and capture-mark-recapture approaches that can be incorporated into routine data collected in large-scale health treatment centres.

There is a suspicion that a number of noncommunicable diseases may be increasing in frequency due to unknown factors in the environment. Unfortunately in many instances, especially for noncancer endpoints and for those that result in death infrequently or after a long latent period, the database does not exist to enable such changes in frequency to be fully assessed. There is therefore a need to set up mechanisms to monitor the occurrence of such diseases. These mechanisms are discussed in more detail below.

THE ROLE OF PUBLIC HEALTH DEPARTMENTS

Many issues in environmental epidemiology arise from the domain of departments of public health. A group of residents near a hazardous waste site, for example, may become concerned with odors from the site; seepage into their grounds or basements; or various symptoms that they may attribute, in the absence of any other information, to the chemicals in the site. These concerns are likely sooner or later to be brought to the notice of the local public health department, usually, at least initially, with requests for reassurance. Further, public health departments are often pressured to study diseases of unknown etiology where there may be insufficient evidence of other causes and often insufficient time from initial exposures for the presumed latent period to be exceeded. The public tend to have unreasonable expectations of what is achievable in epidemiology that may lead to studies that cannot possibly provide an answer to the question.

Unfortunately, most public health departments are ill-equipped to deal with such issues. First, the staff are likely to have little training in either environmental epidemiology or in environmental toxicology, as there are very few such training programs in any country and most public health practitioners will have been trained to cope with other concerns (infectious disease outbreaks, immunisation, maternal and child health, or even cancer, but rarely with any direct relevance to the environment). These deficiencies can lead to simplistic approaches based on invalid assumptions with methods appropriate to infectious disease epidemiology rather than those appropriate to chronic disease. Second, the resources available to such departments for investigation are usually too limited, even if outside specialists capable of performing an appropriate investigation could be found. This may result in a limited investigation being undertaken, which is inconclusive, with a suspicion that if more thorough study had been undertaken a hazard might well have been identified. Third, resources may also be lacking for adequate measurement of exposure. Regulators often do not take measurements relevant to human exposure. Fourth, there may be a politically induced lack of concern about environmental hazards. Such an attitude may derive from economic considerations (the source of possibly hazardous pollutants may bring economic benefits to the area), or from a lack of evidence from other sources that such pollutants can be a concern for public health. Such an attitude may be self-fulfilling if the lack of adequate interest or resources for full investigation in other areas has resulted in an apparently negative finding, which may suggest the inevitability of a further "negative" finding. Fifth, even though it may be clear, as in a chemical spill from a railroad car, that the public has been exposed to a possible hazard, proprietary or "trade secret" concerns may withhold the necessary information on the nature of the exposure from the public health unit so that appropriate remedial action may be delayed. Finally, many exposed populations are too small, the latent period has been too short, the exposure has been too poorly measured, or the outcomes have been too poorly defined, so that at the best a verdict of "not proven" is all that can be expected.

Nevertheless, it seems probable — given the variety of chemicals in hazardous waste sites now or in different localities of the environment, and those which may well be introduced in the future — that the first indication of a hazard from a particular chemical or group of chemicals may still follow an investigation of some event, or cluster, by a public health department. In occupational epidemiology, there are several well-documented instances where a new occupational hazard was first identified by "an alert clinician". Similarly, in environmental epidemiology, there are instances where a hazard was first documented by an appropriate investigation conducted by a public health department. The ability of public health departments to conduct appropriate investigations must therefore be strengthened. Studies initiated by public health departments tend to reflect the circumstances underlying much of the published literature on environmental epidemiology that often results in inconclusive findings, though they may suggest a positive effect. These include the study of small populations, with exposures that are not well characterised. However, if effort were taken to develop databases of such studies, two purposes might be served: it would allow other parties interested in the area to learn what studies are underway, so as to increase sharing of information on study design; and it would eventually allow appropriate combined analysis of the study findings.

EXPOSURE CONSIDERATIONS

Exposure assessment is a crucial and often inadequate component of studies on hazardous waste sites. Exposure assessment entails numerous techniques to measure or estimate the contaminant, its source, the environmental media of exposure, avenues of transport through each medium, chemical and physical transformations, routes of entry to the body, intensity and frequency of contact, and its spatial and temporal concentration patterns. It also includes estimations of total exposure to different compounds and mixtures. Records of ambient pollutant concentrations can sometimes provide a surrogate for exposure, but direct measures of past human exposure have not usually been recorded and must be estimated with models.

The hazardous waste sites potentially responsible for human exposure in the U.S. include mining waste sites, leaking underground storage tanks, pesticide-contaminated sites, federal facilities, radioactive release sites, underground injection wells, municipal gas facilities, unregistered dumps in industrial plants, and wood-preserving plants, as well as known, registered, hazardous waste sites.

Repositories of potentially dangerous substances can thus be found at a number of sites that have been generated by a wide range of activities. However, information about the materials in these sites generally reflects the requirements of environmental engineering and site remediation, rather than public health considerations. Accordingly, whether the materials present in these sites pose a risk cannot readily be determined in the absence of more detailed information about potential human exposure.

The focus of many studies has been on characterising the materials present in a specific site, even though pollutants do not respect such boundaries. Given the potential for movement of materials in ground water and air and the importance of multiple routes of exposure, efforts are needed to estimate plume characteristics to improve the ability to anticipate the movement of pollutants and ultimately to prevent greater human exposure. As a specific example, consider the potential for exposure to pollutants in water. Human exposure comes from recreational use and from water produced for consumption. Such water supplies are derived variously from surface and ground water sources, with variable opportunities for contamination. Further, exposure from domestic water is not limited to ingestion, but includes airborne exposures from materials released during showering, bathing, or cooking.

The best estimate is that ground water is the major source of drinking water for about 50% of the U.S. population. Millions of tons of hazardous materials are slowly migrating into ground water in areas where they could pose problems in the future, even though current risks may be negligible. These include many nonconventional pollutants that are not currently regulated, but which are potentially important hazardous exposures. Preliminary toxicologic studies suggest that nonconventional pollutants have important biologic properties, environmental persistence, and mobility. Additional studies are needed to characterise the mixture of materials deposited as hazardous wastes and to give better estimates of their potential transport and fate in the environment. In the broadest sense, these unidentified, unregulated substances present a risk of unknown magnitude, but the absence of data about them cannot be construed as indicating there is no risk.

DEVELOPING RELEVANT EXPOSURE GRADIENTS

Exposures to synthetic organic chemicals and other modern processes do not occur in neat little packages that can be reliably segregated into media of air, water, or soil. Rather, modern exposure scenarios involve multi-media, multi-temporal levels of many complex chemical compounds. This may be particularly true for the exposures that occur from hazardous waste sites. Further, physical and biologic characteristics of other environmental factors can influence uptake and total dose of chemicals. Thus, heat, humidity, and associated particle size affect the extent of uptake of airborne contaminants, while water hardness, pH, humic acid constituents, and other natural background factors affect exposure to contaminants in water.

In order to link environmental exposures to disease, general types of potentially relevant exposures must be considered:

- Long-standing exposures to well-characterised synthetic organic chemicals, such as pesticides, commercial chemicals, pharmaceutical agents, food, and cosmetic ingredients and additives
- New or previously unassessed exposures such as to industrial materials that reach the general environment either directly or through hazardous waste, including bioengineered products
- Increased UV-B tropospheric radiation from ozone depletion
- Electromagnetic fields and other factors

While much work in the past has relied on exposures classified as "present" or "absent", for some environmental exposures it may not be possible to identify unexposed populations. Efforts must be made to generate mean-ingful gradients of exposures for such populations, including the use of models to improve exposure estimates. These models need to include environmental and biologic fate, population activity patterns, biomonitoring, and biomarkers.

STUDIES IN THE SCIENTIFIC LITERATURE ON HAZARDOUS WASTES IN AIR, WATER, SOIL, AND CONTAMINATED FOOD

Although there is an extensive body of literature on the epidemiology of air pollution, there is little information on airborne exposures from hazardous waste sites. The major difficulties relate to the lack of characterisation of relevant exposures, the small size of the population affected by many expo-sures, and their rapid dilution over large areas. In spite of these difficulties, successful community studies have been done on air pollution patterns from hazardous "point source" exposures. Excesses of angiosarcomas have been noted in residents near a vinyl chloride manufacturing plant (Brady et al., 1977). Another study found increased rates of birth defects in children whose parents lived near such plants (Rosenman et al., 1989). Studies of populations near hazardous waste sites have detected complaints of neurobehavioral symp-toms (Hertzman et al., 1987; Ozonoff et al., 1987). Although recall bias may explain the difference in such subjectively reported symptoms, the possibility exists that such symptoms are more sensitive indicators of exposure than are diseases such as cancer with long latencies.Water is the key medium of concern in exposure from most hazardous waste sites. Episodes of contamination from abandoned sites such as Love Canal, New York and Woburn, Massachusetts have been of particular concern. Evidence on the risk to health from contam-inated water from hazardous waste sites has been largely derived from ecologic studies, and therefore is seldom conclusive as to cause. For example, in New Jersey, studies have linked exposures to hazardous waste sites with increased cancer risks (Najem et al., 1985). A cohort study of residents of North Carolina who drank raw, industrially polluted river water for many years showed a doubling of cancer rates, at times corresponding to the expected latency for cancer (Osborne et al., 1990). Other analytic epidemiology studies, especially as combined in a recent meta-analysis, have shown that trihalomethanes, of concern as contaminants from a number of sites, can increase the risk of bladder and possibly other cancers as contaminants derived from chlorination (Morris et al., 1992). By analogy therefore, contamination of water from hazardous waste sites by such chemicals must also be of concern. For non-cancer endpoints, several reports have found adverse reproductive effects asso-ciated with contaminated drinking water. One study in South Australia found a threefold risk of bearing infants who had defects of the central nervous system among those who drank ground water compared to those who drank rainwater (Dorsch et al., 1984). Persons living in a small valley in Arizona who consumed water contaminated with trichloroethylene were three times more likely to produce offspring with congenital heart disease than those who drank uncontaminated water (Goldberg et al., 1990). A number of other reports have linked consumption of contaminated water with spontaneous abortions, low birthweight or birth defects (National Research Council, 1991). Further, other health effects, including liver and neurological disease, have been associated with waterborne exposure to substances from hazardous waste sites.

Soil provides a usually unrecognised source of exposure to contaminants. Models indicate that adults can be exposed directly, or indirectly through the food chain, and that children incur greater exposures per unit of body weight. However, unless a chemical is extremely potent, the exposure is particularly direct, or there is extensive dust exposure of food or residences, exposure due chiefly to contaminated soil is unusual. Nevertheless, home gardening and ingestion of commercial or recreationally derived fish can be important sources of these contaminants. Mercury contamination of fish is especially prevalent in the Great Lakes region of the U.S. and Canada. Persons who consume fish taken from such waters have average blood levels of DDT and polychlorinated biphenyls (PCBs) several times those found in other population groups (Fiore et al., 1989). PCBs also occur at many hazardous waste sites. Several studies have shown that children exposed to high levels of PCBs prenatally have increased rates of developmental defects (Gladen et al., 1988). Humans may be particularly sensitive to such effects of PCBs.

THE GREY LITERATURE

The grey literature includes state-generated reports and reports on analyses conducted by researchers involved in litigation, some of which contain critical information. Reports on a substantial number of relevant studies have not yet appeared in the conventional peer-reviewed literature. Some of the studies provide useful information to communities, public health officials, and researchers about specific exposures in local areas and general health problems potentially associated with exposures from hazardous waste sites. However, many had been initiated as a response to public concerns with inadequate funding and limitations in exposure assessment and design that undermined their value, so that for scientific reasons they may never be published in the peer-reviewed literature. Others have had their findings sealed by the courts, and thus they cannot be published or reviewed.

THE NEED FOR MORE GENERAL MONITORING

Because of increasing exposures in the general environment, in terms of the overall public good, it seems probable that environmental epidemiology issues will become of increasing importance to public health. Monitoring systems utilising existing or refined routinely collected data sources should be designed to be capable of evaluating the "macro" view of public health, and indeed if environmental contamination continues and becomes more general, these systems may be the only way to determine the extent to which disease rates have changed as a result of such contamination.

The design of such monitoring systems has to take note of the problems that occur in determining the contribution of environmental factors to diseases where few of the causal factors are known. These problems include:

- Difficulties in exposure identification
- Follow-up latency
- Size of the affected population
- Imprecise symptomatology for some of the conditions

This implies that the monitoring systems have to be large, i.e., cover a substantial population; long-term, i.e., continuous and long-running; capable of combination with other similar data systems, i.e., collect data in a systematic standardised fashion; and capable of utilising record linkage to other data sources (e.g., of exposure), i.e., adequate personal identifiers have to be collected and retained with adequate preservation of confidentiality.

Monitoring systems can be newly established (i.e., a special registry set up for a specific purpose as, for example, has been the case for some cancer registries) or, more efficiently, can rely to a greater or lesser extent on existing data systems.

The options include:

- Special surveys (unlikely to be sufficient, because of the requirement for long-term monitoring, though surveys can be repeated periodically)
- Disease reporting systems (as for infectious diseases)
- Capture-mark-recapture systems
- Projects that link existing disease and exposure registries
- Special surveillance mechanisms based on Health Maintenance Organisations
- Special record linkage systems, such as that pioneered in Oxford, U.K. or currently under development in Manitoba, Canada.

For many of the requirements for which routine monitoring is desirable, it would not be possible to set up a special system; or for that matter it is not currently feasible to contemplate extensive universal record linkage systems, even if they were carefully planned solely for the purposes of compiling statistical, research, and disease monitoring data. Hence we need to utilise more efficiently the existing data systems or set up in the cheapest but most efficient way a different type of mechanism.

THE ROLE OF THE RESEARCHER

The role of the researcher in evaluating aggregate statistics, in order to detect trends or patterns that are not apparent at a local level, is critical. For example, an investigation into what was initially thought to be laboratory drift in blood lead concentrations in the Second National Health and Nutrition Examination Survey was eventually found to be the result of a 37% decrease in the population mean blood lead level caused by the fall in the use of lead in gasoline (Annest et al., 1983). This 37% decrease nationwide was of enormous public health significance, and was used to justify the further elimination of lead from gasoline. Yet precisely because it was nationwide, it was not noticed in any geographic comparisons.

An additional role in terms of overall public health depends on the alert researcher. The 2.5-fold increase in mortality in the London smog episode in 1952 could not be determined from hospital data at the time, but became apparent when an investigator compiled weekly mortality data. Similarly, hospital admissions for asthma in children were cut in half in the Utah valley when a steel mill closed down, and returned to the previous level when it reopened. This was also not detected by clinicians or the state health department, but required the examination of hospitalisation data by an alert investigator (Pope, 1991).

Small relative changes are also not likely to be detected unless they are specifically sought. However, small relative changes in the most widespread chronic diseases are of enormous public health consequence. The total mortality consequences of small changes in cardiovascular mortality could be far greater than the effects of localised environmental toxicants. If environmental factors were to be associated with such changes, it would require that those who routinely use aggregate data on such illnesses to be more aware of the potential effect of changes in exposure to environmental factors that should be evaluated. Ecologic analyses are still of value in environmental epidemiology, perhaps particularly in monitoring the effects of unusual or not fully specified exposures, such as from hazardous waste sites. New mapping techniques should be explored to evaluate further the effect of these exposures. It is increasingly recognised that measurement error affects epidemiology studies and statistical techniques are being developed that may reduce its influence.

Issues relevant to trade secrecy, enshrined in legislation designed to protect the interests of commerce rather than public health, can create major problems for public health departments when they are faced with a potential disaster in the form of chemical contamination of the environment. In occupational health, workers' "right to know" the nature and potential hazards from chemicals to which they are exposed has been mandated by legislation in some countries. There are good reasons for extending this principle to chemicals to which the public may be unwittingly exposed.

Critical to improving the research agenda will be the collection of more sophisticated but also more relevant exposure information, both at a population and at an individual level.

There is also a need to more carefully consider the endpoints relevant to the hazards to which the population has been, or will be exposed. Such endpoints not only need to be relevant in providing a possible "early warning" of future increases in disease rates, but also need to be sensitive to our understanding of the natural history of the relevant disease process and of appropriate sensitivity and specificity in terms of carefully evaluated population norms. There is also a possibility that a number of diseases, so far regarded as of unknown etiology, may be affected by environmental exposures. Such a possibility has been raised for insulin-dependent diabetes mellitus (Diabetes Epidemiology Research International, 1987) and may well apply also to other conditions. There is increasing interest in reproductive endpoints relevant to males as well as females; and the need to study other "non-traditional" endpoints, such as those related to the neurological and renal systems, is being recognised as well.

Finally, there is a need to stress the requirements for achieving good practices in environmental epidemiology, not only for academic researchers, but also for those working in public health, industry, and government. Researchers should be regarded as having the ultimate responsibility for the conduct of their studies rather than the organization for which they work. Adoption of guidance for professional practices in environmental epidemiology research might go some way to overcome some of the deficiencies identified in the grey literature discussed above. In addition, to ensure the integrity of the scientific process, epidemiologists should be protected from inappropriate constraints that may be imposed by sponsors or the agencies within which they work. For instance, the guidelines for good epidemiology practices of the Chemical Manufacturers Association Epidemiology Task Force (CMA, 1991) would allow industry to influence or control some aspects of epidemiology research, such as the release of results. This is inappropriate.

However, good epidemiologic research depends on more factors than a set of guidelines for practice, no matter how well conceived. It rests on the existence of investigators with adequate training and experience to design the appropriate study, along with adequate resources to perform the study. Further, the field of environmental epidemiology is advancing in concert with other epidemiology fields as new techniques for exposure assessment, new sources of routinely collected data, and new statistical methodology become available. Hence, what may be regarded as good epidemiology practice now may become outmoded in the future, and may, under certain circumstances, have to be modified as professional judgement dictates.

CONCLUSIONS

Whether the Superfund and other hazardous waste programs protect human health is a critical question with respect to federal and state efforts to clean up hazardous wastes in the U.S. as well as in other countries. To answer this question requires information on the scope of human exposures to contaminants from hazardous waste sites and on the health effects that could be associated with such exposures. The committee was unable to answer this question from its review of the published literature (National Research Council, 1991). Although billions of dollars have been spent during the past decade to study and manage hazardous waste sites in the U.S., an insignificant portion has been devoted to evaluate attendant health risks. The committee was concerned that populations may be at risk that have not been adequately identified, because of the inadequate program of site identification and assessment.

In spite of the limitations of epidemiologic studies of hazardous waste sites, several investigations at specific sites have documented a variety of symptoms of ill health in exposed persons, including low birthweight, cardiac anomalies, headache, fatigue, and neurobehavioral problems. It is less clear whether outcomes with a long delay between exposure and disease have also occurred; however, some studies have detected excesses of cancer in residents exposed to compounds known to occur in hazardous waste sites. Until better evidence is developed, prudent public policy demands that a margin of safety be provided regarding potential health risks from exposures to substances from hazardous waste sites.

REFERENCES

Annest, J. L., Pirkle, J. L., Makuc, D., Neese, J. W., Bayse, D. D., and Kovar, M. G., Chronological trend in blood lead levels between 1976 and 1980. *N. Eng. J. Med.* 308, 1373, 1983.

ATSDR biennial report to congress: October 17, 1986–September 30, 1988. Atlanta: Agency for Toxic Substances and Disease Registry, 1989.

Brady, J., Liberatore, F., Harper, P., Greenwald, P., Burnett, W., Davies, J. N. P., Bishop, M., Polan, A., and Vianna, N., Angiosarcoma of the liver: an epidemiologic survey. *J. Natl. Cancer Inst.* 59, 1383, 1977.

CMA (Chemical Manufacturers Association). Guidelines for Good Epidemiology Practices for Occupational and Environmental Epidemiologic Research. Washington, D.C.: Chemical Manufacturers Association, 1991. [Also available in *J. Occup. Med.* 33, 1221, 1991.]

Diabetes Epidemiology Research International. Preventing insulin dependent diabetes mellitus: the environmental challenge. *Br. Med. J.* 295, 479, 1987.

Dorsch, M. M., Skragg, R. K. R., McMichael, A. J., Baghurst, P. A., and Dyer, K. F., Congenital malformations and maternal drinking water supply in rural South Australia: a case-control study. *Am. J. Epidemiol.* 119, 473, 1984.

Fiore, B. J., Anderson, H. A., Hanrahan, L. P., Olson, L. J., and Sonzongi, W. C., Sport fish consumption and body burdens of chlorinated hydrocarbons: a study of Wisconsin anglers. *Arch. Environ. Health* 44, 82, 1989.

GAO (U.S. Government Accounting Office). Superfund: Public Health Assessments Incomplete and of Questionable Value. Report to the Chairman, Subcommittee on Oversight and Investigations, Committee on Energy and Commerce, House of Representatives. GAO/RCED-91-178, 1991.

Gladen, B. C., Rogan, W. J., Hardy, P., Thullen, J., Tinglestad, J., and Tully, M., Development after exposure to polychlorinated biphenyls and dichlorodiphenyl dichloroethene transplacentally and through human milk. *J. Pediatr.* 113, 991, 1988.

Goldberg, S. J., Lebowitz, M. D., Graver, E. J., and Hicks, S., An association of human congenital cardiac malformations and drinking water contaminants. *J. Am. Coll. Cardiol.* 16, 155, 1990.

Hertzman, C., Hayes, M., Singer, J., and Highland, J., Upper Ottawa street landfill site health study. *Environ. Health Perspect.* 75, 173, 1987.

Morris, R. D., Audet, A. M., Angelillo, I. F., Clamers, T. C., and Mosteller, F., Chlorination, chlorination by-products, and cancer. A meta-analysis. *Am. J. Public Health* 82, 955, 1992.

Najem, G. R., Louria, D. B., Lavenhar, M. A., and Feuerman, M., Clusters of cancer mortality in New Jersey municipalities; with special reference to chemical toxic waste disposal sites and per capita income. *Int. J. Epidemiol.* 14, 528, 1985.

National Research Council. Environmental epidemiology. Volume 1. Public Health and Hazardous Wastes. Washington, D.C., National Academy Press, 1991.

National Research Council. Environmental epidemiology. Volume 2. Data Gaps, Resource Needs and Research Opportunities, 1995.

Osborne, J. S. III, Shy, C. M., and Kaplan, B. H., Epidemiologic analysis of a reported cancer cluster in a small rural population. *Am. J. Epidemiol.* 132 (Suppl 1), S87, 1990.

OTA (U.S. Congress Office of Technology Assessment). Coming clean: Superfund's problems can be solved. OTA-ITE-433. Washington, D.C., U.S. Government Printing Office, 1989.

Ozonoff, D., Colten, M. E., Cupples, A., Heeren, T., Schatzkin, A., Mangione, T., Dresner, M., and Colton, T., Health problems reported by residents of a neighborhood contaminated by a hazardous waste facility. *Am. J. Ind. Med.* 11, 581, 1987.

Pope, C. A., Respiratory hospital admissions associated with $PM^1/_2$ 10, pollution in Utah, Salt Lake, and Cache valleys. *Arch. Environ. Health* 46, 90, 1991.

Rosenman, K. D., Rizzo, J. E., Conomos, M. G., and Halpin, G. J., Central nervous system malformations in relation to two polyvinyl chloride production facilities. *Arch. Environ. Health* 44, 279, 1989.

World Bank. World Development Report, 1992: Development and the Environment. Oxford: Oxford University Press, 1992.

EPIDEMIOLOGIC RESEARCH PRIORITIES ON THE HEALTH RISK OF DRINKING WATER CONTAMINANTS

5

Kenneth P. Cantor

CONTENTS

EPIDEMIOLOGIC EVIDENCE

In this chapter, we review the evidence linking drinking water contaminants with human disease, and suggest areas for further research. Our primary emphasis is on cancer; however, we also briefly discuss adverse pregnancy outcomes and current issues in waterborne infectious disease.

Cancer

Although drinking water quality in industrialised nations is generally considered to be excellent, there is concern that waterborne contaminants may contribute to the human cancer burden. Limited evidence from epidemiologic studies suggests a link of several classes of contaminants with elevated cancer risk. This review summarises the epidemiologic evidence addressing human cancer risk from several classes of drinking water contaminants: (1) asbestiform particles; (2) radionuclides; (3) inorganic solutes, especially arsenic and nitrate; and (4) synthetic organic chemicals. These groups frequently occur in combination and their interactive effects may differ from the sum of individual effects. This applies to their carcinogenic potential, the difficulty and validity of epidemiologic investigations, and potential control measures.

Asbestos

Asbestos fibers (primarily chrysotile, amosite, and crocidolite) enter drinking water primarily through weathering from natural deposits such as serpentine; by release from asbestos-cement pipes; and from processes associated with mining and production of iron ore, such as those which contaminated Lake Superior and the water supply of Duluth, Minnesota through the mid-1970s.[1,2] The carcinogenic potential of asbestos is influenced by the size, shape, and crystalline structure of the fibers; and these are modified by physicochemical processes resulting from exposure to water or gastric fluid.[3]

Despite evidence from occupational studies of the carcinogenicity of asbestos in the human lung, pleura, and gastrointestinal tract, epidemiologic studies of populations served by water containing high concentrations of asbestos have failed to yield conclusive results. All but one study involving asbestos in drinking water are ecologic. Two studies in census tracts of the San Francisco Bay area found associations between measured (naturally occurring) asbestos in drinking water and incidence of cancer of the oesophagus, stomach, pancreas in both sexes,[4] lung (males), and female gallbladder and peritoneum.[5] Factors such as diet, smoking, and occupation could not be adequately controlled. In Quebec communities, asbestos in drinking water was associated with mortality of cancers of the stomach (males), pancreas (females), and lung (males).[6] An ecologic study[7] and a case-control study[8] in Washington State (U.S.), based on incident cancer, did not find overall patterns of association with consuming drinking water from a river containing high levels of naturally occurring asbestos. However, positive associations for male stomach cancer, based on small numbers, were observed in the case-control study. This may be of importance because others have found associations for this site. Positive associations were also found for male pharyngeal cancer. Duluth, Minnesota had high levels of asbestos in its drinking water from 1955 to 1973. Cancer mortality[9] and incidence[10,11] in Duluth or the county in which it is located were compared to rates of other Minnesota cities (or their respective counties). Some excesses of gastrointestinal mortality and morbidity were observed, but patterns were inconsistent.

Cancer incidence as related to drinking water distributed by asbestos-cement water mains has been evaluated in Connecticut,[12,13] Utah,[14] and Woodstock, New York[15] and mortality has been evaluated in Escambria County, Florida,[16] with inconsistent findings. In Utah, an association was found for male kidney cancer and female leukaemia, and in Woodstock, New York, cancer of the oral cavity.

Radionuclides

Traces of natural and man-made radioactivity from radionuclides are found in drinking water supplies throughout the world. Levels vary geographically with local soil and rock conditions and may be increased by industrial and other point discharges. Monitoring has been used to estimate the occurrence of uranium,[17] radon-222,[18] radium,[19] and total radioactivity;[20] and the contribution of drinking water to total natural background radiation.[18,21,22]

Waterborne radon is responsible for most of the population dose of alpha radiation from drinking water, with the primary exposure being to the lungs via airborne radon released from water.[22] Ingested radon or other radionuclides are thought to be less important as environmental determinants of cancer.[23] The predominant source of indoor Rn-222 in houses is the soil underlying and adjacent to the foundation. Generally, water contributes less than 2% of total household radon. However, in unusual circumstances, ground water is the predominant source of indoor airborne radon.[24] The few investigations of cancer and radioactivity in drinking water are ecologic in design. A study of county leukaemia incidence rates in Florida found a relative risk of 1.5 for total leukaemia and 2.0 for acute myeloid leukaemia in high ground water radium vs. low radium counties.[25] A study of county cancer mortality in Maine found an association between female rates for 1950–1969 and average county radon concentrations in water.[24] In Iowa, the incidence of cancers of the lung and bladder among males and of the lung and breast among females was elevated in towns with a radium-226 level in the water supply exceeding 5.0 pCi/l.[26] A later study in 59 Iowa towns revealed a small, increasing trend for total leukaemia incidence with radium content in drinking water consistent with either no or a small effect.[27] Although several disaggregate studies have examined the association of airborne radon in homes with lung cancer risk, the contribution of radon in household drinking water to the total has not been considered separately.

Inorganic Solutes

Several inorganic solutes commonly found in drinking water, among them arsenic and nitrate, are suspected to increase cancer risk in exposed populations. Although there is no credible evidence linking fluoride ion to cancer risk, it is included here because of broad population exposure, the existence of an extensive epidemiologic data base, and equivocal evidence from animal studies suggesting a carcinogenic risk.

Arsenic. Among studies of metals and transition elements in drinking water, the strongest evidence of a link to human cancer risk comes from studies of arsenic. Initial investigations reported elevated prevalence of skin cancer in areas of chronic arsenism from contaminated water supplies in Mexico,[28] Argentina,[29] Taiwan,[30,31] and Chile.[32] However, skin cancer prevalence and/or incidence has not been elevated in other places where populations are also exposed to elevated arsenic levels in drinking water,[33,34] suggesting that other cofactors may be important.

Further study in Taiwan revealed geographic associations of arsenic in drinking water with risk of mortality from other cancers, including cancers of the bladder, kidney, lung, nasal cavity, liver in both sexes, and prostate.[35-37] The strongest correlations were found for bladder and kidney cancers. In a case-comparison study of bladder, lung, and liver cancer in an area with arsenic-contaminated artesian wells, Chen and co-workers[38] found positive dose-response relationships, with odds ratios of 3.9 for bladder cancer, 3.4 for lung cancer, and 2.7 for liver cancer among those who had used contaminated water for 40 years or longer, as compared with never-users. Limited support for the bladder cancer findings comes from an observation of excess bladder cancer mortality among patients treated with Fowler's solution (potassium arsenate).[39]

Nitrate.   Nitrate ion occurs in surface and ground waters in concentrations ranging from less than 1.0 mg/l to over 100 mg/l. When drinking water nitrate nitrogen is well below 10 mg/l (the WHO recommended maximum limit), most ingested nitrate comes from dietary sources, and averages about 100 mg/day. However, when water levels are near or exceed this level, water may increase nitrate intake to 200 mg/day or more.[40]

Nitrate can act as a procarcinogen, interacting with secondary amines and amides to form *N*-nitroso compounds which are powerful carcinogens in many species of laboratory animals,[41] after reduction in the saliva of nitrate to nitrite.[42-47]

The direct epidemiologic evidence for carcinogenic effects is equivocal. Some studies, most of them of ecologic design, have shown associations between nitrate concentrations in drinking water and gastric cancer. Brain cancer is also of concern, although there is less evidence. Positive geographic associations of nitrate with gastric cancer have been found in Chile, Hungary, England, Columbia, the U.S., Italy, and Denmark. However, other studies have not demonstrated a positive association. Geographic associations of this type may be subject to "publication bias", whereby positive associations may be more likely to find their way into print than null findings. Retrospective cohort studies of fertiliser workers who are presumably exposed to nitrate dusts do not show excess cancer risk.[48-50] A case-control study of gastric cancer deaths from Wisconsin did not show a link with nitrate level of the water source at the last residence of decedents.[51]

Fluoride.   Fluoride is present in most natural waters, at levels usually less than 1.0 mg/l, and rarely exceeding 5.0 gm/l. Its natural source is mainly the dissolution of fluoride ion from minerals, such as apatite, amphibole, and fluorite. A principal source in an increasing number of communities over the past four decades has been the successful prophylactic addition of fluoride to prevent dental caries. The usual dosage is in the range 0.7–1.2 mg/l.

An early study suggested that the overall cancer mortality rates of the ten largest U.S. cities that practiced water fluoridation were significantly higher than those of the ten largest that did not.[52] The differences disappeared when relevant sociodemographic variables were adequately controlled.[53,54] Additional studies, almost all of them of ecologic design, provided no supporting evidence. The epidemiologic findings on fluoridation and cancer risk have been independently reviewed by an international panel[55] and by three separate expert committees convened in the U.S.[56,57] and Great Britain.[58] These groups concluded that the available evidence does not support the hypothesis that fluoride in drinking water supplies influences cancer risk. A recent laboratory finding of "equivocal evidence" of osteosarcoma excesses among male rats (but not female rats, or male or female mice) after lifetime ingestion of 79 ppm fluoride in drinking water has fueled recent concern.[59] In the brief period since, epidemiologic assessments have found no time trend or geographic pattern of bone cancer or osteosarcoma consistent with a causal role for fluoride in drinking water.[60-63]

Organic Chemicals

The increasing ability to detect low levels of contaminants have both increased our knowledge of organics in drinking water, and greatly complicated risk assessment, since detected levels of compounds are often quite low, beneath the concentrations at which health effects detectable by epidemiologic methods may be possible. Hundreds of organic chemicals have been found in drinking waters, but most occur infrequently and at low concentration (below or at the part per billion range). At least 40 have been characterised as known or suspected carcinogens, including vinyl chloride, benzene, and chloromethyl ether.[64]

Organic Chemicals of Industrial, Agricultural, Commercial, and Domestic Origin. Contamination of underground and surface water with organic chemicals from industrial, agricultural, commercial, and domestic sources, as well as from hazardous waste disposal sites, is increasingly found. Systematic epidemiologic study has been hampered by difficulties in estimating the levels, timing, and specific chemicals involved in past exposures; the relatively small populations usually exposed to high contaminant levels; and the problem of deciding which health endpoints, or intermediate biological markers, to examine. When effects are observed, it is often impossible to determine the specific exposures.

Many studies have examined county and local cancer mortality and/or incidence rates in places with hazardous waste sites, or places with contaminated municipal water supplies. In a nationwide study in the U.S., age-adjusted cancer mortality rates from 339 counties with hazardous waste sites were compared with rates from 2726 other counties.[65] Significant associations were found for lung, bladder, stomach, large intestine, and rectal cancers in white males and females; oesophagus in white males; and breast in white females. In the state of New Jersey, female (but not male) leukaemia incidence rates in 27 towns were associated with an index of volatile organic chemicals in municipal drinking water.[66] In a Pennsylvania county with a major toxic waste disposal site, bladder cancer mortality among white males was significantly elevated.[67] In Massachusetts, exposure to tetrachloroethylene leachate from improperly cured vinyl-lined distribution pipes was associated with nonsignificant risk elevations for bladder cancer and leukaemia.[68]

An assessment of cancer incidence rates for the period 1955-1977 in the vicinity of the Love Canal disposal site in Niagara Falls, New York revealed elevated lung cancer that was not consistent across age groups.[69] Other cancers did not appear to be elevated, but the statistical power to detect elevated rates of less common sites was limited. Rates of chromosomal aberrations and sister chromatid exchange frequencies were as expected.[70] A Finnish community with drinking water contaminated with chlorophenols, probably from saw-mills, had an elevated incidence of soft-tissue sarcoma and non-Hodgkin's lymphoma.[71] These tumours have been linked with exposure to the closely related chlorinated phenoxyacetic acids and/or their dioxin contaminants.[72] Liver cell cancer in China was strongly linked to consuming drinking water from ditches highly polluted with agricultural runoff that presumably con-tained a variety of organic and other chemicals.[73]

A cluster of childhood leukaemia cases associated with contaminated community drinking water in Woburn, Massachusetts, has been the subject of scientific, legal, and political controversy.[74-77] Drinking water contaminants included trichloroethylene (267 ppb), tetrachloroethylene (21 ppb), trichloro-trifluoroethane (23 ppb), and dichloroethylene (28 ppb). Elevated levels of 22 metals were found in 61 test wells drilled to sample ground water.[74] Lympho-cyte abnormalities were noted among family members of cases.[77]

Organic Chemicals: Disinfection By-Products. Chemical by-products are found in almost every disinfected drinking water supply. Both chlorine and ozone combine with organics in drinking water, with chlorinated by-products thought to be the more toxic. Most epidemiologic studies have focused on populations exposed to water treated with chlorine, the most common disin-fectant used in the U.S. Among chlorination by-products, the volatile triha-lomethanes (THM) account for 20–80% of the covalently bound halogen. Included among higher molecular weight non-volatiles are a large number of carboxylic acids, aldehydes, ketones, and ethers.[78,79] The concentration of total THM ranges from less than 1 ppb (in treated water from deep wells low in organics) to several hundred parts per billion (certain chlorinated surface waters). The contrast between the relatively high concentration of chlorination by-products in treated surface waters, as compared with well waters, has served as the basis for epidemiologic evaluation.

Ecologic studies of mortality or incidence were used first to estimate the potential scope of the problem. In many studies, the county was the geographic unit of observation. Age-adjusted and site-, sex-, and race-specific county cancer mortality rates were used as outcomes; and characteristics (chlorinated vs. nonchlorinated, surface vs. ground, THM level) of the predominant county drinking water source as the exposure variable.[80-84] Incidence rates were used as outcomes in studies of water quality and cancer in Iowa towns,[85,86] Norwe-gian municipalities,[87] and Finnish cities.[88] The sites of cancer most commonly associated with surface water, chlorination, or THM level were the urinary bladder, colon, and rectum.[89,90]

The second tier of studies were of case-comparison design, and used mor-tality records as the sources of case and comparison subjects. In many,[91-95] the exposure variable was a characteristic of the water supply (such as sur-face/ground, etc.) that served the decedent's last residence, as abstracted from death certificate records. In later studies, information about previous residences and sources of drinking water was also used.[96-98] Results were largely supportive of the findings from the earlier ecologic studies in showing elevated risk for colon, rectal, and bladder cancers. A notable exception is the finding of Lawrence et al.[97] of no association between type of water source or imputed past THM level with colorectal cancer.

The third set of investigations were disaggregate studies in which past exposures were estimated through linking historical community water supply records with residential history information gathered in personal interviews. One community cohort follow-up study and four case-comparison studies used this approach.[99-103] Three of the four case-comparison studies also gathered information about the tap water consumption of individuals. In Washington County, Maryland, Wilkins and Comstock[99] found elevated (but not statistically significant) bladder cancer incidence among men, and cancer of the liver among women in the drinking water subcohort supplied with chlorinated surface water at home as compared with people with a history of unchlorinated ground water use.

Colon cancer was the subject of case-comparison interview studies in North Carolina[102] and Wisconsin.[101] In the former, an association was observed with use of chlorinated surface water, but only among cases more than 60 years old. In Wisconsin, where the authors estimated trihalomethane ingestion at various times in the past, no associations with colon cancer were noted.

Bladder cancer risk as related to water source and tap water consumption was evaluated in a case-control interview study of 3000 cases and 6000 controls conducted in ten areas of the U.S.[100] Bladder cancer risk increased with the amount of tap water consumed, and the association was stronger among respondents who resided for 40–59 or 60+ years at places served by surface water than for fewer years. There was no increase of risk with tap water consumption among persons who had lived at places served by non-chlorinated ground water for most of their lives. Similar findings were found in a much smaller case-control study of bladder cancer and disinfection methods in Colorado.[103]

Adverse Pregnancy Outcomes

Several studies have looked at associations between drinking water contaminants and adverse outcomes of pregnancy, including low birthweight, spontaneous abortion, intrauterine growth retardation, and several types of congenital malformations. In many studies, the exposure to drinking water contaminants was poorly characterised, or consisted of a mixture of many contaminants, making interpretation difficult. Two studies were conducted in populations that were also evaluated for cancer induction after exposure to contaminated water. An excess of low birthweight babies was observed among offspring of Love Canal residents, especially in the most highly exposed swale area during the period of active dumping.[104] In Woburn, Massachusetts, with exposure to contaminated ground water, positive associations were found for perinatal deaths, eye/ear anomalies, and CNS/chromosomal/oral cleft anomalies; however, many congenital anomalies were evaluated and no associations were noted with most.[74]

In rural South Australia, women who principally consumed ground water had a statistically significant increased risk of bearing a malformed child (relative risk [RR] = 2.8), especially of CNS and musculoskeletal systems.[105] Nitrate was elevated in the ground water, and a seasonal gradient in risk was evident among ground water consumers. Nitrate, a common ground water contaminant, has not been linked with birth anomalies in other locations, so it may be serving here as an indicator compound.

A cluster of adverse pregnancy outcomes in Santa Clara County, California, including spontaneous abortions (odds ratio [OR] = 2.3, confidence interval [CI] = 1.3–4.2) and congenital malformations (OR = 3.1, CI = 1.1–10.4), was first thought to be related to consumption of water known to be contaminated with trichloroethylene.[106] However, detailed evaluation was not consistent with a causal relationship.[107-109] In Tucson Valley, Arizona, congenital cardiac malformations were linked to consumption of ground water contaminated with industrial chemicals, especially trichloroethylene, dichloroethylene, and chromium.[110] In an ecologic study from Iowa,[111] analysis of birth certificate data and water utility records found an association of chloroform level with intrauterine growth retardation (<5th percentile), but not with low birthweight (<2,500 gm) or prematurity (<37 weeks gestation).

Infectious Disease

Control of waterborne infectious disease in industrial societies is largely attributed to drinking water and wastewater treatment. Most infectious outbreaks linked to contaminated water can be traced to breakdowns in treatment systems. However, usual chlorination practice is not adequate to kill all *Giardia* cysts, or apparently to inactivate many viruses. Outbreaks of *Giardia* have been reported from places with chlorinated water.[112,113] In a randomised trial in Montreal, in which the study populations drank chlorinated filtered and unfiltered municipal water, higher gastrointestinal disease rates occurred in the latter group.[114]

RESEARCH NEEDS

Epidemiologic evaluation of drinking water contaminants is dependent on the accidental or inadvertent exposure of human populations to potentially toxic materials. Clearly, exposures known to be dangerous, on the basis of experimental or human data, must be reduced or eliminated as rapidly as possible. However, there are many situations where control of exposure cannot be justified given current knowledge of the potential hazard. In these situations, focused research is needed to determine the presence and magnitude of risk to health.

With regard to exposure conditions, drinking water contaminants can be grouped into two broad classes: (1) chemicals in water which are localised in their distribution and often occur at high concentration, and (2) contaminants which have widespread occurrence. These categories are not rigid, and some types of pollutants clearly occur in both situations. However, the distinction may be helpful in designing epidemiologic studies and in developing conceptual models of public health risk. The major sources of localised water contaminants are toxic waste disposal sites or industrial complexes. The chemical classes include chlorinated organic solvents and other industrial chemicals and production intermediates, including other organics and metals. Among contaminants with more widespread distribution are by-products of chlorination (including chloroform and other trihalomethanes), naturally occurring asbestiform particles, and radon. Arsenic compounds, nitrate, and some pesticides can occur at high concentrations both locally and at lower levels with widespread distribution.

Cancer Risk and Research Opportunities

Asbestos

The evidence from epidemiologic studies to date is not conclusive. Asbestiform particles are widespread in the aqueous environment, and a period of 20–40 years may be required before their effects in humans are seen. Cancer rates in high-exposure areas should receive continued surveillance. In addition, asbestos in drinking water should be considered as a risk factor whenever large case-control studies of stomach, kidney, pancreas, and other cancers are conducted in regions with some sources of asbestos-contaminated water.

Radionuclides

As discussed earlier, drinking water is usually a minor contributor to total airborne domestic radon, typically accounting for about 2% of the total. Radiation doses from ingested radon are minor. However, in places where drinking water may contribute a substantial fraction of airborne radon in homes, careful measurement should be made of its contribution, since techniques for reducing radon in water may differ from mitigation of the airborne gas. Several ongoing studies of lung cancer and airborne radon in houses are addressing the issue of risk via this exposure route. In addition, radium in water deserves further evaluation, and this exposure should be considered in case-control studies of leukaemia and of lung cancer.

Inorganic Solutes

Arsenic.   Recent studies from Taiwan strengthen the evidence that arsenic causes human cancer of several sites by ingestion of ambient trace amounts. The broad distribution of these metals in drinking water indicates a need for further study. Additional disaggregate investigations in unstudied populations, which exploit situations of high exposure or high disease incidence, should receive top priority. The anatomic sites of greatest interest are bladder, kidney, lung, nasal cavity, and liver. A standard case-control or population-based cohort approach should be used in investigating this question.

Nitrate.   The direct epidemiologic evidence for a link between nitrate exposure and cancer is weak. However, the carcinogenic potential of nitrate in drinking water is suggested by recent studies that demonstrate the endogenous production of N-nitrosoproline following ingestion of waterborne nitrate and proline.[46,47] The relationship of nitrate in water with cancer may be complex and depend on other cofactors, including host characteristics, diet, and environmental factors. Further work on the mechanism of action is needed, especially on conditions that influence formation of N-nitroso compounds. Progress in the direct epidemiologic evaluation of a cancer-nitrate link is hampered by limitations in our ability to estimate past exposure to nitrate, especially waterborne exposures that may have occurred many decades ago. Further work is called for in this area of exposure assessment. In light of the widespread and increasing contamination of water supplies with nitrate, especially in agricultural regions, and the demonstrated potential for N-nitroso compound formation, research into the health effects of nitrate deserves high priority. In studying nitrate, the contribution of diet must be considered, both with regard to total nitrate as well as micronutrients such as vitamin C that can inhibit the N-nitrosation reaction.

Fluoride.   After extensive evaluation in ecologic studies, the evidence for a fluoride-cancer link in human populations is very weak indeed. However, recent laboratory findings of "equivocal evidence" for osteosarcoma in male rats fed high levels of fluoride[59] have raised the question of a possible link in people for risk of this cancer. The few completed studies do not support the hypothesis. However, given the widespread nature of the exposure, further evaluation of osteosarcoma in case-control studies is needed.

Other Inorganic Solutes.   There is little evidence regarding the carcinogenic risk posed by other inorganic solutes in drinking water. Several metals may be carcinogenic in occupational settings, such as chromium and nickel, and effects of these in human populations deserve evaluation when they occur in water. Study designs will depend on the particular circumstance.

Synthetic Organic Chemicals

Organic Chemicals of Industrial, Agricultural, Commercial, and Domestic Origin. Among water contaminants, carcinogenic effects of commercial organic chemicals in human populations are perhaps the most challenging to evaluate. Exposed populations are often small, and exposures transient. Estimating past exposure is especially challenging, even in relatively uncomplicated situations. Several studies suggest a possible a link between leukaemia and chlorinated hydrocarbon solvents in water, but support from occupational studies is lacking, after allowing for the relative exposure levels in the two settings. The Finnish study showing excess soft tissue sarcoma and non-Hodgkin's lymphoma after exposure to waterborne chlorophenols[71] is of special interest, but the exposure circumstance was unusual, and replication may not be possible. Continued health surveillance of populations with known exposures is of high priority. In addition, studies of biological markers of exposure and of effect are recommended.

Disinfection By-Products. There is mounting evidence for a link between exposure to chlorination by-products, especially at levels found in many chlorinated surface waters, and cancers of the bladder, rectum, and possibly colon. However, the existing data base is weak, and additional studies are needed to resolve the uncertainties and to develop more accurate estimates of risk. Given the widespread exposure involved and the suggestions of risk that are available, further work deserves high priority. Careful replication of case-control studies of these sites, with inclusion of detailed lifetime exposure histories, collection of information on dietary and fluid ingestion histories, and data on other risk factors, is recommended. In addition, there is a need for metabolic and bio-marker studies that can provide evidence about possible mechanisms of action of commonly found chlorination by-products.

Adverse Pregnancy Outcomes and Drinking Water Contaminants

The very limited database on drinking water contaminants and adverse outcomes of pregnancy carries some limited suggestion of risk following exposure to chlorinated volatile hydrocarbons. Possible effects include intrauterine growth retardation after chloroform exposure, and congenital cardiac and other malformations after exposure to a mixture of chlorinated solvents in water. The preliminary observation of intrauterine growth retardation and chloroform exposure in Iowa[111] appears to be testable in other settings, because the exposure is widespread and well documented. Other associations may be more difficult to evaluate, because exposures are more sporadic and poorly documented. However, where excessive exposure has been documented and good birth records are available, such study may be possible. The type of study design chosen will depend on the circumstances of exposure.

Infectious Disease

For many years, public health authorities believed that current wastewater and drinking water treatment was fully adequate to protect against all water-borne infectious disease. There is a thus need to expand on recent observations suggesting that a third or more of cases of endemic gastrointestinal disease, resulting in severe diarrhoea and vomiting, are due to low titers of virus or other factors resistant to chlorine disinfection.

RESEARCH PRIORITIES

The research needs outlined above are grouped here in high, moderate, or low priority categories. Priorities are primarily based on extent of exposure, evidence of hazard, and our view of the potential for productive research. The cancer sites listed are for guidance only.

High priority
1. Nitrate (cancer sites: stomach, brain, others)
2. Disinfection by-products, especially chlorination by-products (cancer sites: bladder, colon, rectum; outcomes of pregnancy)
3. Arsenic (cancer sites: skin, bladder, kidney, liver, others)
4. Leachate from toxic waste dumps and industrial sites, including chlorinated solvents, metals, and other industrial chemicals (cancer types: leukaemia, others; outcomes of pregnancy)

Moderate priority
1. Infectious agents
2. Fluoride (cancer site: osteosarcoma)

Low priority
1. Asbestos
2. Radon and other radionuclides

REFERENCES

1. Millette, J. R., Clark, P. J., Stober, J., and Rosenthal, M., Asbestos in water supplies of the U.S.. *Environ. Health Perspect.* 53, 45, 1983.
2. Langer, A. M., Maggiore, C. M., Nicholson, W. J., Rohl, A. N., Rubin, I. B., and Selikoff, I. J. ,The contamination of Lake Superior with amphibole gangue minerals. *Ann. N.Y. Acad. Sci.* 330, 549, 1979.
3. Seshan, K., How are the physical and chemical properties of chrysotile asbestos altered by a 10-year residence in water and up to 5 days in simulated stomach acid? *Environ. Health Perspect.* 53, 143, 1983.

4. Conforti, P. M., Kanarek, M. S., Jackson, L. A., Cooper, R. C., and Murchio, J. C., Asbestos in drinking water and cancer in the San Francisco Bay area: 1969–1974 incidence. *J. Chron. Dis.* 34, 211, 1981.

5. Kanarek, M. S., Conforti, P. M., Jackson, L. A., Cooper, R. C., and Murchio, J. C., Asbestos in drinking water and cancer incidence in the San Francisco Bay area. *Am. J. Epidemiol.* 112, 54, 1980.

6. Wigle, D. T., Cancer mortality in relation to asbestos in municipal water supplies. *Arch. Environ. Health.* 32, 185, 1977.

7. Polissar, L., Severson, R. K., Boatman, E. S. and Thomas, D. B., Cancer incidence in relation to asbestos in drinking water in the Puget Sound region. *Am. J. Epidemiol.* 116, 314, 1982.

8. Polissar, L., Severson, R. K., and Boatman, E. S., A case-control study of asbestos in drinking water and cancer risk. *Am. J. Epidemiol.* 119, 456, 1984.

9. Mason, T. J., McKay, F. W., and Miller, R. W., Asbestos-like fibers in Duluth water supply. *JAMA* 228, 1019, 1974.

10. Levy, B. S., Sigurdson, E., Mandel, J., Laudon, E., and Pearson, J., Investigating possible effects of asbestos in city water: Surveillance of gastrointestinal cancer incidence in Duluth, Minnesota. *Am. J. Epidemiol.* 103, 362, 1976.

11. Sigurdson, E. E., Levy, B. S., Mandel, J., McHugh, R., Michienzi, L. J., Jagger, H., and Pearson, J., Cancer morbidity investigations: Lessons from the Duluth study of possible effects of asbestos in drinking water. *Environ. Res.* 25, 50, 1981.

12. Meigs, J. W., Walter, S. D., Heston, J. F., Millette, J. R., Craun, G. F., Woodhull, R. S., and Flannery, J. T., Asbestos cement pipe and cancer in Connecticut 1955-1974. *J. Environ. Health* 42, 187, 1980.

13. Harrington, J. M., Craun, G. F., Meigs, J. W., Landrigan, P. J., Flannery, J. T., and Woodhull, R. S., An investigation of the use of asbestos cement pipe for public water supply and the incidence of gastrointestinal cancer in Connecticut, 1935-1973. *Am. J. Epidemiol.* 107, 96, 1978.

14. Sadler, T. D., Rom, W. N., Lyon, J. L., and Mason, J. O., The use of asbestos-cement pipe for public water supply and the incidence of cancer in selected communities in Utah. *J. Community Health* 9, 285, 1984.

15. Howe, H. L., Wolfgang, P. E., Burnett, W. S., Nasca, P. C., and Youngblood, L., Cancer incidence following exposure to drinking water with asbestos leachate. *Public Health Rep.* 104, 251, 1989.

16. Millette, J. R., Craun, G. F., Stober, J. A., Kraemer, D. F., Tousignant, H. G., Hildago, E., Duboise, L., and Benedict, J., Epidemiology study of the use of asbestos-cement pipe for the distribution of drinking water in Escambia County, Florida. *Environ. Health Perspect.* 53, 91, 1983.

17. Cothern, D. R. and Lappenbusch, W. L., Occurrence of uranium in drinking water in the U.S. *Health Phys.* 45, 89, 1983.

18. Cross, F. T., Harley, N. H., and Hofmann, W., Health effects and risks from radon-222 in drinking water. *Health Phys.* 48, 649, 1985.

19. Mays, C. W., Rowland, R. E., and Stehney, A. F., Cancer risk from the lifetime intake of Ra and U isotopes. *Health Phys.* 48, 635, 1985.

20. Hess, C. T., Michel, J., Horton, T. R., Prichard, H. M., and Coniglio, W. A., The occurrence of radioactivity in public water supplies in the U.S.. *Health Phys.* 48, 553, 1985.

21. Cothern, C. R., Lappenbusch, W. L., and Michel, J., Drinking-water contribution to natural background radiation. *Health Phys.* 50, 33, 1986.
22. Nazaroff, W. W., Doyle, S. M., Nero, A. V., and Sextro, R. G., Potable water as a source of airborne 222Rn in U.S. dwellings: A review and assessment. *Health Phys.* 52, 281, 1987.
23. National Research Council, *Drinking Water and Health. Volume 1*, Washington, D.C.:National Academy of Sciences, 1977.
24. Hess, C. T., Weiffenbach, C. V., and Norton, S. A., Environmental radon and cancer correlations in Maine. *Health Phys.* 45, 339, 1983.
25. Lyman, G. H., Lyman, C. G., and Johnson, W., Association of leukemia with radium ground water contamination. *JAMA* 254, 621, 1985.
26. Bean, J. A., Isacson, P., Hahne, R. M. A., and Kohler, J., Drinking water and cancer incidence in Iowa. II. Radioactivity in drinking water. *Am. J. Epidemiol.* 116, 924, 1982.
27. Fuortes, L., McNutt, L. A., and Lynch, C., Leukemia incidence and radioactivity in drinking water in 59 Iowa towns. *Am. J. Public Health* 80, 1261, 1990.
28. Cebrian, M. E., Albores, A., Aguilar, M., and Blakely, E., Chronic arsenic poisoning in the north of Mexico. *Human Toxicol.* 2, 121, 1983.
29. Bergoglio, R. M., Mortalidad por cancer en zonas de aguas arsenicales de la Provincia de Cordoba, Republica Argentina. *Prensa Med. Argent.* 51, 994, 1964.
30. Tseng, W. P., Chu, H. M., and How, S. W., Prevalence of skin cancer in an endemic area of chronic arsenicism in Taiwan. *J. Natl. Cancer Inst.* 40, 453, 1968.
31. Tseng, W. P., Effects and dose-response relationships of skin cancer and blackfoot disease with arsenic. *Environ. Health Perspect.* 19, 109, 1977.
32. Borgono, J. M. and Grieber, R., Estudio epidemiologico del arsenicismo en la ciudad de Antofagasta. *Rev. Med. Chil.* 99, 702, 1971.
33. Morton, W., Starr, G., Pohl, D., Stoner, J., Wagner, S., and Weswig, P., Skin cancer and water arsenic in Lane County, Oregon. *Cancer* 37, 2523, 1976.
34. Harrington, J. M., Middaugh, J. P., Morse, D. I., and Housworth, J., A survey of a population exposed to high concentrations of arsenic in well water in Fairbanks, Alaska. *Am. J. Epidemiol.* 108, 377, 1978.
35. Wu, M. M., Kuo, T. L., Hwang, Y. H., and Chen, C. J., Dose-response relation between arsenic concentration in well water and mortality from cancers and vascular diseases. *Am. J. Epidemiol.* 130, 1123, 1989.
36. Chen, C. J., Chuang, Y. C., Lin, T. M., and Wu, H. Y., Malignant neoplasms among residents of a blackfoot disease-endemic area in Taiwan: High-arsenic artesian well water and cancers. *Cancer Res.* 45, 5895, 1985.
37. Chen, C. J. and Wang, C. J., Ecological correlation between arsenic level in well water and age-adjusted mortality from malignant neoplasms. *Cancer Res.* 50, 5470, 1990.
38. Chen, C. J., Chuang, Y. C., You, S. L., Lin, T. M., and Wu, H. Y., A retrospective study on malignant neoplasms of bladder, lung and liver in blackfoot disease endemic area. *Br. J. Cancer* 53, 399, 1986.
39. Cuzick, J., Sasieni, P., and Evans, S., Ingested arsenic, keratoses, and bladder cancer. *Am. J. Epidemiol.* 136, 417, 1992.

40. Chilvers, C., Inskip, H., Caygill, C., Bartholomew, B., Fraser, P., and Hill, M., A survey of dietary nitrate in well-water users. *Int. J. Epidemiol.* 13, 324, 1984.

41. International Agency for Research on Cancer, *IARC Monographs on the Evaluation of the Carcinogenic Risk of Chemicals to Humans, Volume 17: Some N-Nitroso Compounds*, Lyon, France: IARC, 1978.

42. Eisenbrand, G., Speigelhalder, B., and Preussmann, R., Nitrate and nitrite in saliva. *Oncology* 37, 227, 1980.

43. Walters, C. L. and Smith, P. L. R., The effect of water-borne nitrate on salivary nitrite. *Food Cosmet. Toxicol.* 19, 297, 1981.

44. Walters, C. L., The exposure of humans to nitrite. *Oncology* 37, 289, 1980.

45. Hart, R. J. and Walters, C. L., The formation of nitrite and N-nitroso compounds in salivas in vitro and in vivo. *Food Cosmet. Toxicol.* 21, 749, 1983.

46. Moller, H., Landt, J., Pedersen, E., Jensen, P., Autrup, H., and Jensen, O. M., Endogenous nitrosation in relation to nitrate exposure from drinking water and diet in a Danish rural population. *Cancer Res.* 49, 3117, 1989.

47. Mirvish, S. S., Grandjean, A. C., Moller, H., Fike, S., Maynard, T., Jones, L., Rosinsky, S., and Nie, G. N., Nitrosoproline excretion by rural Nebraskans drinking water of varied nitrate content. *Cancer Epidemiol. Biomark Prev.* 1, 455, 1992.

48. Rafnsson, V. and Gunnarsdottir, H., Mortality study of fertiliser manufacturers in Iceland. *Br. J. Ind. Med.* 47, 721, 1990.

49. Al-Dabbagh, S., Forman, D., Bryson, D., Stratton, I., and Doll, R., Mortality of nitrate fertiliser workers. *Br. J. Ind. Med.* 43, 507, 1986.

50. Fraser, P., Chilvers, C., Day, M., and Goldblatt, P., Further results from a census based mortality study of fertiliser manufacturers. *Br. J. Ind. Med.* 46, 38, 1989.

51. Rademacher, J. J., Young, T. B., and Kanarek, M. S., Gastric cancer mortality and nitrate levels in Wisconsin drinking water. *Arch. Environ. Health* 47, 292, 1992.

52. Burk, D. and Yiamouyiannis, J., Fluoridation and cancer. *Congressional Rec.* July 21, 1975.

53. Hoover, R. N., McKay, F. W., and Fraumeni, J. F., Fluoridated drinking water and the occurrence of cancer. *J. Natl. Cancer Inst.* 57, 757, 1976.

54. Chilvers, C., Cancer mortality and fluoridation of water supplies in 35 U.S. cities. *Int. J. Epidemiol.* 12, 397, 1983.

55. International Agency for Research on Cancer, Inorganic fluorides. In: *IARC Monographs on the Evaluation of Carcinogenic Risk of Chemicals to Humans, Volume 27: Some Aromatic Amines, Anthraquinones and Nitroso Compounds, and Inorganic Fluorides*, Lyon, France: IARC, 1982, 235.

56. National Research Council, Inorganic solutes. In: *Drinking Water and Health. Volume 1*, Washington, D.C.: National Academy of Sciences, 1977, 369.

57. U.S. Public Health Service. Committee to Coordinate Environmental Health and Related Programs, Subcommittee on Fluoride *Review of Fluoride: Benefits and Risks*, Washington, D.C.:Department of Health and Human Services, 1991.

58. Knox, E. G., *Fluoridation of Water and Cancer: A Review of the Epidemiological Evidence*, London: Report of a Working Party: Her Majesty's Stationary Office, 1985.

59. Bucher, J. R., Hejtmancik, M. R., Toft, J. D., II, Persing, R. L., Eustis, S. L., and Haseman, J. K., Results and conclusions of the National Toxicology Program's rodent carcinogenicity studies with sodium fluoride. *Int. J. Cancer* 48, 733, 1991.

60. McGuire, S. M., Vanable, E. D., McGuire, M. H., Buckwalter, J. A., and Douglass, C. W., Is there a link between fluoridated water and osteosarcoma? *J Am. Dent. Assoc.* 122, 38, 1991.

61. Hoover, R. N., Devesa, S., Cantor, K., and Fraumeni, J. F., Jr., Appendix F. Time Trends for Bone and Joint Cancers and Osteosarcomas in the Surveillance, Epidemiology and End Results (SEER) Program, National Cancer Institute. In: Review of Fluoride: Benefits and Risks. Report of the Ad Hoc Subcommittee on Fluoride of the Committee to Coordinate Environmental Health and Related Programs, Washington, D.C.: Public Health Service, DHHS, 1991,

62. Hrudey, S.E., Soskolne, C. L., Berkel, J., and Fincham, S., Drinking water fluoridation and osteosarcoma. *Can. J. Public Health* 81, 415, 1990.

63. Mahoney, M. C., Nasca, P. C., Burnett, W. S., and Melius, J. M., Bone cancer incidence rates in New York State: Time trends and fluoridated drinking water. *Am. J. Public Health* 81, 475, 1991.

64. International Agency for Research on Cancer, *Overall Evaluations of Carcinogenicity: An Updating of IARC Monographs Volumes 1 to 42*, Lyon, France: IARC, 1987.

65. Griffith, J., Duncan, R. C., Riggan, W. B., and Pellom, A. C., Cancer mortality in U.S. counties with hazardous waste sites and ground water pollution. *Arch. Environ. Health* 44, 69, 1989.

66. Fagliano, J., Berry, M., Bove, F., and Burke, T., Drinking water contamination and the incidence of leukemia: An ecologic study. *Am. J. Public Health* 80, 1209, 1990.

67. Budnick, L. D., Sokal, D. C., Falk, H., Logue, J. N., and Fox, J. M., Cancer and birth defects near the Drake Superfund site, Pennsylvania. *Arch. Environ. Health* 39, 409, 1984.

68. Aschengrau, A., Ozonoff, D., Paulu, C., Coogan, P., Vezina, R., Heeren, T., and Zhang, Y., Cancer risk and tetrachloroethylene (PCE) contaminated drinking water in Massachusetts. *Arch. Environ. Health* 48, 284, 1993.

69. Janerich, D. T., Burnett, W. S., Feck, G., Hoff, M., Nasca, P., Polednak, A. P., Greenwald, P., and Vianna, N., Cancer incidence in the Love Canal area. *Science* 212, 1404, 1981.

70. Heath, C. W., Nadel, M. R., Zack, M. M., Chen, A. T. L., Bender, M. A., and Preston, R. J., Cytogenetic findings in persons living near the Love Canal. *JAMA* 251, 1437, 1984.

71. Lampi, P., Hakulinen, T., Luostarinen, T., Pukkala, E., and Teppo, L., Cancer incidence following chlorophenol exposure in a community in southern Finland. *Arch. Environ. Health* 47, 167, 1992.

72. Lilienfeld, D. E. and Gallo, M., A. 2,4 D, 2,4,5 T, and 2,3,7,8 TCDD: An Overview. In: *Epidemiologic Reviews Volume 11*, edited by Armenian, H. K., Gordis, L., Gregg, M. B., and Levine, M. M., Baltimore, pp. 28-58: *American Journal of Epidemiology*, 1989.

73. Delong, S., Drinking water and liver cell cancer. *Chin. Med. J.* 92, 748, 1979.

74. Lagakos, S. W., Wessen, B. J., and Zelen, M., An analysis of contaminated well water and health effects in Woburn, Massachusetts. *J. Am. Stat. Assoc.* 81, 583, 1986.

75. MacMahon, B., Prentice, R. L., Rogan, W. J., Swan, S. H., Robins, J. M., and Whittemore, A. S., Comments and rejoinder on Lagakos, Wessen, and Zelen article on contaminated well water and health effects in Woburn, Massachusetts. *J. Am. Stat. Assoc.* 81, 597, 1986.

76. Cutler, J. J., Parker, G. S., Rosen, S., Prenney, B., Healey, R., and Caldwell, G. G., Childhood leukemia in Woburn, Massachusetts. *Public Health Rep.* 101, 201, 1986.

77. Byers, V. S., Levin, A. S., Ozonoff, D. M., and Baldwin, R. W., Association between clinical symptoms and lymphocyte abnormalities in a population with chronic domestic exposure to industrial solvent-contaminated domestic water supply and a high incidence of leukaemia. *Cancer Immunol. Immunother.* 27, 77, 1988.

78. Stevens, A. A., Moore, L. A., and Miltner, R. J., Formation and control of non-trihalomethane disinfection by-products. *J. Am. Water Works Assoc.* 81;8, 54, 1989.

79 Singer, P. C. and Chang, S. D., Correlations between trihalomethanes and total organic halides formed during water treatment. *J. Am. Water Works Assoc.* 81;8, 61, 1989.

80. Cantor, K. P., Hoover, R., Mason, T. J., and McCabe, L. J., Associations of cancer mortality with halomethanes in drinking water. *J. Natl. Cancer Inst.* 61, 979, 1978.

81. Hogan, M. D., Chi, P. Y., and Hoel, D. G., Association between chloroform levels in finished drinking water supplies and various site-specific cancer mortality rates. *J. Environ. Pathol. Toxicol.* 2, 873, 1979.

82. Kuzma, R. J., Kuzma, C. M., and Buncher, C. R., Ohio drinking water source and cancer rates. *Am. J. Public Health* 67, 725, 1977.

83. Page, T., Harris, R. H., and Epstein, S. S., Drinking water and cancer mortality in Louisiana. *Science* 193, 55, 1976.

84. Salg, J., Cancer Mortality Rates and Drinking Water in 346 Counties of the Ohio River Valley Basin, University of North Carolina, Chapel Hill: Ph.D. Thesis, 1977.

85. Bean, J. A., Isacson, P., Hausler, W. J., and Kohler, J., Drinking water and cancer incidence in Iowa. I. Trends and incidence by source of drinking water and size of municipality. *Am. J. Epidemiol.* 116, 912, 1982.

86. Isacson, P., Bean, J. A., and Lynch, C., Relationship of cancer incidence rates in Iowa municipalities to chlorination status of drinking water. In: *Water Chlorination: Environmental Impact and Health Effects Volume 4*, edited by Jolley, R.L., Brungs, W.A., Cotruvo, J.A., Cumming, R.B., Mattice, J.S., and Jacobs, V.A. Ann Arbor, MI: Ann Arbor Science, 1983, 1353.

87. Flaten, T. P., Chlorination of drinking water and cancer incidence in Norway. *Int. J. Epidemiol.* 21, 6, 1992.

88. Koivusalo, M., Jaakkola, J., Vartiainen, T., Hakulinen, T., Karjalainen, S., Pukkala, E., and Tuomisto, J., Drinking water mutagenicity and gastrointestinal and urinary tract cancers: An ecological study in Finland. *Am. J. Public Health* 84, 1223, 1994.

89. Wilkins, J. R., III, Reiches, N. A., and Kruse, C. W., Organic chemical contaminants in drinking water and cancer. *Am. J. Epidemiol.* 110, 420, 1979.

90. National Research Council, Epidemiological studies. In: *Drinking Water and Health. Volume 3*, edited by Safe Drinking Water Committee, Washington, D.C.: National Academy Press, 1980, 5.

91. Alavanja, M., Goldstein, I., and Susser, M., A case control study of gastrointestinal and urinary tract cancer mortality and drinking water chlorination. In: *Water Chlorination: Environmental Impact and Health Effects Volume 2*, edited by Jolley, R.L., Gorchev, H., and Hamilton, D.H., Jr. Ann Arbor, MI: Ann Arbor Science, 1978, 395.

92. Young, T. B., Kanarek, M. S., and Tsiatis, A. A., Epidemiologic study of drinking water chlorination and Wisconsin female cancer mortality. *J. Natl. Cancer Inst.* 67, 1191, 1981.

93. Brenniman, G. R., Vasilomanolakis-Lagos, J., Amsel, J., Namekata, T., and Wolff, A. H., Case-control study of cancer deaths in Illinois communities served by chlorinated or nonchlorinated water. In: *Water Chlorination: Environmental Impact and Health Effects, Volume 3*, edited by Jolley, R.L., Brungs, W.A., and Cumming, R.B. Ann Arbor, MI: Ann Arbor Science, 1980, 1043.

94. Gottlieb, M. S. and Carr, J. K., Case-control cancer mortality study and chlorination of drinking water in Louisiana. *Environ. Health Perspect.* 46, 169, 1982.

95. Crump, K. S. and Guess, H. A., Drinking water and cancer: Review of recent epidemiological findings and assessment of risks. *Annu. Rev. Public Health* 3, 339, 1982.

96. Gottlieb, M. S., Carr, J. K., and Morris, D. T., Cancer and drinking water in Louisiana: Colon and rectum. *Int. J. Epidemiol.* 10, 117, 1981.

97. Lawrence, C. E., Taylor, P. R., Trock, B. J., and Reilly, A. A., Trihalomethanes in drinking water and human colorectal cancer. *J. Natl. Cancer Inst.* 72, 563, 1984.

98. Zierler, S., Feingold, L., Danley, R. A., and Craun, G., Bladder cancer in Massachusetts related to chlorinated and chloraminated drinking water: A case-control study. *Arch. Environ. Health* 43, 195, 1988.

99. Wilkins, J. R., III and Comstock, G. W., Source of drinking water at home and site-specific cancer incidence in Washington County, Maryland. *Am. J. Epidemiol.* 114, 178, 1981.

100. Cantor, K. P., Hoover, R., Hartge, P., Mason, T. J., Silverman, D. T., Altman, R., Austin, D. F., Child, M. A., Key, C. R., Marrett, L. D., Myers, M. H., Narayana, A. S., Levin, L. I., Sullivan, J. W., Swanson, G. M., Thomas, D. B., and West, D. W., Bladder cancer, drinking water source, and tap water consumption: A case-control study. *J. Natl. Cancer Inst.* 79, 1269, 1987.

101. Young, T. B., Wolf, D. A., and Kanarek, M. S., Case-control study of colon cancer and drinking water trihalomethanes in Wisconsin. *Int. J. Epidemiol.* 16, 190, 1987.

102. Cragle, D. L., Shy, C. M., Struba, R. J., and Siff, E. J., A case-control study of colon cancer and water chlorination in North Carolina. In: *Water Chlorination: Chemistry, Environmental Impact and Health Effects. Volume 5*, edited by Jolley, R.L., Bull, R.J., Davis, W.P., Katz, S., Roberts, M.H. Jr., and Jacobs, V.A. Chelsea, MI: Lewis Publishers, Inc., 1985, 153.

103. McGeehin, M. A., Reif, J. S., Becker, J., and Mangione, E., A case-control study of bladder cancer and water disinfection in Colorado. *Am. J. Epidemiol.* 138, 492, 1993.

104. Vianna, N. J. and Polan, A. K., Incidence of low birthweight among Love Canal residents. *Science* 226, 1217, 1984.

105. Dorsch, M. M., Scragg, R. K. R., McMichael, A. J., Baghurst, P. A., and Dyer, K. F., Congenital malformations and maternal drinking water supply in rural South Australia: A case-control study. *Am. J. Epidemiol.* 119, 473, 1984.

106. Deane, M., Swan, S. H., Harris, J. A., Epstein, D. M., and Neutra, R. R., Adverse pregnancy outcomes in relation to water contamination, Santa Clara County, California, 1980-1981. *Am. J. Epidemiol.* 129, 894, 1989.

107. Shaw, G. M., Swan, S. H., Harris, J. A., and Malcoe, L. H., Maternal water consumption during pregnancy and congenital cardiac anomalies. *Epidemiology* 1, 206, 1990.

108. Swan, S. H., Shaw, G., Harris, J. A., and Neutra, R. R., Congenital cardiac anomalies in relation to water contamination, Santa Clara County, California, 1981-1983. *Am. J. Epidemiol.* 129, 885, 1989.

109. Wrensch, M., Swan, S., Lipscomb, J., Epstein, D., Fenster, L., Claxton, K., Murphy, P. J., Shusterman, D., and Neutra, R., Pregnancy outcomes in women potentially exposed to solvent-contaminated drinking water in San Jose, California. *Am. J. Epidemiol.* 131, 283, 1990.

110. Goldberg, S. J., Lebowitz, M. D., Graver, E. J., and Hicks, S., An association of human congenital cardiac malformations and drinking water contaminants. *J. Am. Coll. Cardiol.* 16, 155, 1990.

111. Kramer, M. D., Lynch, C. F., Isacson, P., and Hanson, J. W., The association of waterborne chloroform with intrauterine growth retardation. *Epidemiology* 3, 407, 1992.

112. Kent, G. P., Greenspan, J. R., Herndon, J. L., Mofenson, L. M., Harris, J. A. S., Eng, T. R., and Waskin, H. A., Epidemic giardiasis caused by a contaminated public water supply. *Am. J. Public Health* 78, 139, 1988.

113. Craun, G. F., Waterborne giardiasis in the U.S. 1965-84. *Lancet* 2, 513, 1986.

114. Payment, P., Richardson, L., Siemiatycki, J., Dewar, R., Edwardes, M., and Franco, E., A randomized trial to evaluate the risk of gastrointestinal disease due to consumption of drinking water meeting current microbiological standards. *Am. J. Public Health* 81, 703, 1991.

6

PESTICIDES AND CANCER: STATUS AND PRIORITIES

Aaron Blair and Shelia H. Zahm

CONTENTS

INTRODUCTION

With their introduction in the 1940s, synthetic pesticides have become an increasingly important weapon in the attempt to control pests and disease vectors. The benefits of pesticides cannot be denied, but there may also be health risks. Consideration of possible long-term health effects arose with the environmental movement.[1] The intense public debate regarding the risks and benefits of pesticides has involved numerous groups including manufacturers and formulators, farmers, commercial pesticide applicators, urban users (for pest control in gardens and homes), and the general public (who may have involuntary exposures from food, air, and water). Controversial scientific issues also surround pesticide evaluation including mechanisms of carcinogenicity, relationship of animal bioassays to human risk assessment, dose-response relationships, potential for thresholds, and single vs. multiple exposures.

Until the 1980s, most toxicologic information on the carcinogenicity of pesticides was provided by animal bioassays.[2] They provide considerable evidence that some pesticides may be hazardous to human health. A summary of findings on over 40 pesticides tested in the National Toxicology Program in the U.S. found no evidence of carcinogenicity in mice or rats for 19 pesticides (40% of those tested); 6 (13%) were carcinogenic in both sexes of both species, 10 (21%) were carcinogenic in both sexes of one species, and 6 (13%) were carcinogenic in one sex in one species.[3] The 16 pesticides carcinogenic in both sexes of at least one species included organochlorine and organophosphate insecticides, herbicides, fungicides, and fumigants, suggesting that no chemical class of pesticides can be considered problem free.

EPIDEMIOLOGIC EVIDENCE

Few epidemiologic investigations on the health effects of pesticides were conducted until the 1980s. This is surprising given the early evidence of their potential hazard in bioassays. The limited epidemiologic effort may have been due to a belief that effects of individual pesticides could not be investigated because of the multiplicity of human exposures. That belief is still widely held today.

Cancer from pesticide exposure has received more attention than other potential health outcomes, but concerns regarding neurologic, reproductive, immunologic, and developmental outcomes are growing. Human studies focusing on non-cancer outcomes are badly needed. The collection of papers edited by Baker and Wilkinson provides a thorough review of a variety of disease outcomes, including cancer.[4] This review will focus on cancer, but recommendations usually also apply to other outcomes.

TYPES OF EPIDEMIOLOGIC STUDIES AVAILABLE

Reviews limited to cancer have focused on farmers,[5-7] herbicides,[8,9] or pesticides in general.[2,3,10-13] Both case-control and cohort designs have been employed in the evaluation of pesticides and cancer. Numerous case-control studies have focused on leukaemia, lymphoma, soft-tissue sarcoma, and childhood cancer, with fewer investigations on other sites. Both case-control and cohort designs have tended to focus on risks associated with general pesticide exposure. In many studies, exposure assessment was based entirely upon the occupational titles with no attempt to identify the specific chemicals used. Fewer have attempted to link cancer with classes of pesticides, or with specific chemicals (Table 1). Examples of studies evaluating pesticides in general are studies of farmers[5,6,14-25] and applicators.[26-34] Many case-control studies have focused on occupational groups with potential exposure to pesticides (see the reviews[2,5,7,10,12]). Other investigations have focused on classes of pesticides,[35-59] many for phenoxyacetic acid herbicides. A growing number of studies are available that have focused on specific pesticides.[60-82] Specific pesticides evaluated include 2,4-D, 2,4,5-T, dioxin, DDT, 2, MCPA, aldrin, dieldrin, endrin, DBCP, PBB, chlordane, heptachlor, ethylene dibromide, arsenic, atrazine, and others.

Table 1   Selected Epidemiologic Studies of Pesticides and Cancer by Level of Exposure Detail

Exposure Category	References
By Occupation	
Farmers	Blair et al. (5,6a,6b), Burmeister (14), Saftlas (15a), Wiklund et al. (16), Reif et al. (17), Ortega (18), Stark et al. (19), Brownson et al. (20), Delzell, Grufferman (21), Gallagher et al. (22), Rafnsson, Gunnarsdottir (23), Carlson, Petersen (24), Almas, Odegard (25)
Applicators/ manufacturers	Blair et al. (26), Barthel (27), MacMahon et al. (28), Wicklund et al. (29), Corrao et al. (30), Alavanja et al. (31), Alberghini et al. (32), Sathiakumar et al. (33), Cantor, Booze (34)
By Pesticide Class	Hardell, Sandstrom (35), Hardell et al. (36), Donna et al. (37), Smith et al. (38), Eriksson et al. (39), Hardell et al. (40), Lynge (41), Pearce et al. (42), Vineis et al. (43), Woods et al. (44), Hardell, Eriksson (45), Donna et al. (46), Eriksson et al. (47), Wingren et al. (48), Wigle et al. (49a), Morrison et al. (49b), LaVecchia et al. (50), Coggon et al. (51), Vineis et al. (52), Green (53), Persson et al. (54), Boffetta et al. (55), Cantor, Blair (56), Morris et al. (57), Flodin et al. (58), Williams et al. (59)
By Specific Pesticide	Ribbens (60), Hearn et al. (61), Wong et al. (62), Ditraglia et al. (63), Ott et al. (64), Wang, MacMahon (65), Mabuchi et al. (66), Austin et al. (67), Saracci et al. (68), Fingerhut et al. (69), Cantor et al. (70), Riihimaki et al. (71), Axelson et al. (72) Bond et al. (73), Hoar et al. (74), Coggon et al. (75), Ott et al. (76), Zahm et al. (77a-77c), Brown et al. (78a,78b), Zahm et al. (79), Garabrant et al. (80), Eriksson, Karlsson (81), Falck et al. (82a), Wolf et al. (82b)

The International Agency for Research on Cancer (IARC) has performed evaluations for a number of pesticides.[83,84] The IARC has concluded that there is at least limited evidence for human carcinogenicity for the following pesticides: amitrole, chlordane/heptachlor, chlorophenols, creosotes, DDT, ethylene dibromide, ethylene oxide, mirex, toxaphene, and occupational exposure to insecticides.

Summarising the large volume of epidemiologic work on pesticides is difficult because inconsistencies occur from study to study and because many studies do not focus on specific chemicals. Among the more frequently investigated cancers, however, some patterns appear (Table 2). Among groups with potential exposure to pesticides, relative risks greater than one tend to occur for leukaemia; non-Hodgkin's lymphoma; multiple myeloma; and cancers of the lung, bladder, and brain.[2] Relative risks are often not large and some reports show deficits. Inconsistencies are not surprising, however, given the failure in most studies to provide a detailed assessment of exposures. Although few investigations have provided information by level or duration of pesticide contact, exposure-response gradients have been observed for lung cancer among structural pesticide applicators,[26,27] lung cancer and DDT,[67] pancreatic cancer and DDT,[80] soft-tissue sarcoma and phenoxyacetic acid herbicides or dioxins,[35,39,69] and non-Hodgkin's lymphoma and phenoxyacetic acid herbicides.[36,49,50,54,74,77]

Table 2   Pesticides Classes Associated with Cancer Among Humans Based on Epidemiologic Research

Pesticide Class	Cancers
Phenoxyacetic acid herbicides	Lymphoma, soft tissue sarcoma
Organochlorine insecticides	Leukaemia, lymphoma, soft tissue sarcoma, pancreas, breast, lung, neuroblastoma
Organophosphate insecticides	Lymphoma, leukaemia
Arsenical insecticides	Lung, skin
Triazine herbicides	Ovary

In studies of applicators in Germany[27] and the U.S.,[26] the risk of lung cancer rose to nearly 3-fold among those employed for 20 or more years. These workers may have used many pesticides, but the time periods of the studies and type of treatment suggest that organochlorine insecticides (i.e., DDT, chlordane, heptachlor) predominated. An association between lung cancer and DDT is also suggested by a non-significant exposure-response relationship between serum blood levels of DDT and subsequent occurrence of lung cancer.[67] Slight excesses of lung cancer have also been reported among manufacturers and applicators exposed to chlordane and heptachlor.[28,65]

Exposure-response patterns have also been reported between phenoxyacetic acid herbicides and soft-tissue sarcoma[35,39,69] and non-Hodgkin's lymphoma.[36,49,50,54,74,77] In studies of non-Hodgkin's lymphoma from the U.S., relative risks were also greater among subjects who did not use protective equipment[74] and among those who delayed in changing to clean clothing,[77] two factors suggesting that heavier exposure results in greater risk.

Associations between ovarian cancer and triazine herbicides;[37,46] breast cancer and insecticides;[82] pancreatic cancer and DDT;[80] leukaemia and crotoxyphos, dichlorvos, famphur, pyrethrins, and methoxychlor;[78] non-Hodgkin's lymphoma and carbaryl, chlordane, diazinon, dichlorvos, lindane, malathion, toxaphene,[70] and organophosphates as a class[77] are each based on only one study.

Given the clear evidence that some pesticides are carcinogenic in laboratory animals, the epidemiologic evidence seems sufficient to indicate that human pesticide exposures are not desirable and are likely to increase the risk of some cancers.

RESEARCH NEEDS

Mechanism of Action

Most bioassays are designed to detect the effects of early stage carcinogens. Some of the carcinogenic pesticides may operate in this fashion. In an evaluation of 65 pesticides by a battery of genotoxic tests, about one-half were found to be active in several of the tests.[85] Some pesticides may also operate through epigenetic mechanisms.[2] For example, mirex is an effect promoter of skin cancer in the mouse,[86] triazines may act through inhibition of pituitary activity,[84,87] and phenoxyacetic acid herbicides may affect immune function.[88] It has also recently been proposed that organophosphates may play a role in oncogenesis through their inhibition of serine esterases, enzymes which are critical components in the cytolytic activities of the T lymphocytes and natural killer cells.[89]

Epidemiologic studies of chronic disease and pesticide exposure have generally not attempted to evaluate mechanism of action. Few have even used time of action in the analyses. In studies where time since first exposure has been considered, risks have generally been larger with longer latency.[26,28,70,74,78,80] These evaluations suggest that the pesticides may be operating at an early stage of carcinogenesis. We are unaware of any attempts to evaluate risks in relation to time since last exposure, an approach that is necessary to assess action at a later stage.

Research is needed to assemble information on mechanisms of pesticide action. Data from completed studies could be analysed by time since last exposure to assess whether pesticides are operating at an early or a late stage in the carcinogenic process. Future investigations should include collection of tumour tissue from cases for evaluation of genetic susceptibility and blood from controls to relate exposure to possible biomarkers for pesticidal action, including immune system function, hormone receptors, point mutations, and chromosomal aberrations.

Exposure and Disease Assessment

Assessment of exposure is typically a major limitation in all epidemiologic studies and this is also true for studies on pesticides.[90] Study of chronic disease risks from exposure to pesticides is complicated by multiple exposures which change over time.[91] However, we believe that for many pesticide exposure situations this problem may have been over-emphasised.[92] In any case, the multiple-exposure situation is equally complex for many other areas of epidemiologic research, for example, diet and disease; yet dramatic progress has been made in assessing diet in epidemiologic studies over the past decade.

Improvements similar to those made for diet can also be accomplished in the evaluation of exposure to pesticides. Because it is unlikely that a single factor which will serve as a "gold standard" for exposure assessment will be found, improvements will require combining information from a variety of sources. Traditional, as well as relatively unusual, approaches will be required. Traditional sources such as work records of, and interviews with, study subjects must be supplemented with information from purchase records, licensing information, collaborative interviews with pesticide suppliers, and environmental and biologic monitoring.

Methodologic studies should be developed to evaluate the reliability and validity of the various approaches to exposure assessment. For example, the reliability of interview information could be assessed by reinterview; information from interviews with applicators should be compared with information from knowledgeable co-workers and with work records, purchase receipts, and diaries; and records and interview data could be compared with environmental and biologic monitoring. Methodologic approaches used by nutritional epidemiologists can serve as a model for this work.

Although we believe that assessment of pesticide exposures can be improved in epidemiologic studies, the effect of misclassification on risk estimates must be considered in interpretation of results. Very often the direction of this effect is incorrectly perceived. Errors in exposure assessment, and disease diagnosis, may be differential or nondifferential.[93] Case-response bias, a form of differential misclassification of exposures, is of particular concern in case-control studies. More accurate reporting of exposure by cases than controls biases risk estimates upward and this would create false positive findings. Although there are few clear-cut examples of case-response bias actually occurring, procedures to detect this bias should be included in case-control studies. Techniques that may be employed include addition of individuals with a disease not related to the exposure of interest (e.g., another cancer); querying subjects about the understanding of the relationship between the disease and exposure of interest; and attempting to corroborate information provided by a sample of the cases and controls through records, interviews with co-workers, or environmental or biologic monitoring. Case-response bias can be entirely avoided in cohort studies where exposure is evaluated before diagnosis of the disease.

Nondifferential misclassification of exposure is a major concern in all epidemiologic studies and, for some diseases, errors in diagnosis are of equal concern. Nondifferential errors are more likely to occur than differential errors and they tend to diminish estimates of risk and dilute exposure-response gradients.[93] Weaknesses and limitations in exposure assessment in most case-control and all cohort studies are more likely to be nondifferential than differential. Small amounts of nondifferential misclassification can result in a major reduction in relative risks[94] with associations more likely to be missed than falsely implicated. Thus, when considering exposure misclassification, it is crucial to determine whether it is likely to be of the differential or nondifferential type because they can have opposing effects on risk estimates.

Diagnosis of diseases that may be associated with pesticides is uneven. For several cancers, diagnosis is quite good, e.g., lung and bladder cancer, but for other cancers of interest, problems may occur. Non-Hodgkin's lymphoma and soft-tissue sarcoma are both collections of cancers of different histologic types and these histologic types may have different etiologies. Inclusion of different histologic types in a study may dilute effects if exposure associations are with specific cell types. It is difficult, however, to develop a study of sufficient size to assess risks by subcategory. When analysis by histologic types is feasible, however, the results may be informative. In a study of leukaemia and pesticide exposure,[78] risks often varied between the different histologic types, with excesses particularly evident for chronic lymphocytic leukaemia. In unpublished data combined from two studies of 2,4-D and non-Hodgkin's lymphoma,[74,77] the risk was elevated for most histologic types, but was exceptionally high for follicular large cell lymphoma (relative risk = 10.0, based on 8 cases). Diagnosis is more difficult for some nonmalignant outcomes, e.g., fertility, immunotoxicity, neurologic lesions, developmental difficulties, than for cancer in epidemiologic studies because of technological and practical difficulties. Many of these outcomes can be well diagnosed only with intensive clinical evaluation, and the absence of disease registries makes this difficult and expensive in epidemiologic investigations.

Overall Effect of Exposure

Exposure-response gradients are critical in establishing causality. Analyses based on a dichotomous (ever vs. never) exposure classification are worthwhile in early, hypothesis-generating investigations. From Table 1, however, it is clear that this area of cancer epidemiology is maturing so that it now requires more sophisticated exposure assessments in order to advance our understanding. Future studies of cancer must include some type of exposure ranking system and, wherever possible, they should have a quantitative component.

Epidemiologic investigations of nonmalignant disease and pesticides are at an earlier stage than those for cancer, and hypothesis-generating studies are still needed. Studies by occupation and pesticide class can be used to sharpen our focus for more analytic efforts.

Subgroups at Special Risk

Subgroups with a special susceptibility have long been a concern with regard to environmental hazards. In the area of pesticides, two groups stand out: children and genetically susceptible subgroups. Results from studies of childhood cancer and pesticide exposures are shown in Table 3. The magnitude of the relative risks from these studies is striking. Risks among children who would have indirect exposures are, in general, greater than among adults who

Table 3   Cancer Risks Among Children Associated
          with Pesticide Exposure

Cancer	Relative Risk	Type of Exposure	Reference
Brain	4.0	Living on a farm	95
Brain	2.4	Parents farmers	96
Leukaemia	3.8	Used pesticides in the home	97
Leukaemia	3.5	Parental use of pesticides	98
Leukaemia	2.7	Parental use of pesticides	99
Bone	6.1	Parental use of pesticides	100

directly handle pesticides. This may point to a special susceptibility of children.

Persons with a family history of cancer may represent another group that may be more susceptible than the public at large. For most cancers, a familial history approximately doubles the risk for an individual. Identification of the genetic component of this familial risk has been a major focus of experimental carcinogenesis. The overall purpose of this line of investigation has been to identify individuals especially susceptible to cancer. This biomarker approach should be incorporated into epidemiologic investigations. Susceptibility can also be evaluated in traditional epidemiologic investigations by comparing risks of cancer from environmental exposures among subjects with and without a family history of cancer. This approach was used in a study of non-Hodgkin's lymphoma and agricultural exposures among women.[77b] Relative risks of non-Hodgkin's lymphoma from insecticide exposure were 0.6 with no family history of cancer, 1.2 with a family history of any type of cancer, 1.5 with a family history of lymphatic or hematopoietic cancer, and 2.3 with a first-degree relative with a lymphatic or hematopoietic cancer. Other reports have noted a similar interaction between smoking and genetic susceptibility on the risk of lung cancer.[101,102] The data above suggest that environmental exposures may be a special problem among persons with a family history of cancer.

Hypotheses to be Tested

Studies that include detailed exposure assessments are needed for the tumours that have been previously associated with pesticides, i.e., leukaemia; non-Hodgkin's lymphoma; multiple myeloma; soft-tissue sarcoma; and cancers of the lung, bladder, and brain. Studies of nonmalignant disease are also urgently needed. Assessments should consider evaluation of indirect exposure from drift, ground water contamination, and food residues; as well as direct exposure from personal use in application on the job, in homes, and on lawns and gardens.

Future epidemiologic investigations should include biomarker components to evaluate mechanisms of action for pesticides, particularly with regard to immunotoxicology.

Epidemiologic studies should include efforts to evaluate cancer risks among potentially susceptible populations. This should include traditional information on family history of cancer, as well as newer biologic markers.

Methodologic work is needed to improve approaches to develop quantitative assessments of pesticide exposures from occupational and general environmental contacts.

REFERENCES

1. Carson, R. I., *Silent Spring*. Houghton Mifflin, Boston, MA, 1962.
2. Blair, A., Axelson, O., Franklin, C., Paynter, O. E., Pearce, N., Stevenson, D., Trosko, J. E., Vainio, H., Williams, G., Woods, J., and Zahm, S. H., Carcinogenic effects of pesticides. In: *The Effects of Pesticides on Human Health,* Baker, S. R., and Wilkinson, C. F. (Eds.). Advances in Modern Toxicology XVIII, Princeton Scientific Publishing Company, Princeton, NJ, 1990, pp. 210-260.
3. Hoover, R. N. and Blair, A., Pesticides and cancer. In: DeVita, V. T. Jr., Hellman, S., and Rosenberg, S. A. (Eds.). *Pesticides and Cancer.* J.B. Lippincott Company, Philadelphia, PA, 1991, pp. 1-11.
4. Baker, S. R. and Wilkinson, C. F. (Eds.). The Effects of Pesticides on Human Health. Adv. in Modern Toxicol. XVIII, Princeton Scientific Publishing Company, Princeton, NJ, 1990.
5. Blair, A., Malker, H., Cantor, K. P., Burmeister, L., and Wiklund, K., Cancer among farmers. A review. *Scand. J. Work Environ. Health* 11:397-407, 1985.
6a. Blair, A. and Zahm, S., Cancer among farmers. In: *Health Hazards of Farming.* Cordes, D.H., and Rea, D.F. (Eds.). Hanley and Belfus, Inc, Philadelphia, PA, 1991. pp. 335-370.
6b. Blair, A., Zahm, S. H., Pearce, N. E., Heineman, E. F., and Fraumeni, J.F. Jr., Clues to cancer etiology from studies of farmers. *Scand. J. Work Environ. Health* 18:209-215, 1992.
7. Pearce, N. and Reif, J. S., Epidemiologic studies of cancer in agricultural workers. *Am. J. Ind. Med.* 18:133-142, 1990.
8a. Johnson, E. S., Association between soft tissue sarcomas, malignant lymphomas, and phenoxy herbicides/chlorophenols: Evidence from occupational cohort studies. *Fundam. Appl. Toxicol.* 14:219-234, 1990.
8b. Morrison, H. I., Wilkins, K., Semenciw, R., Mao, Y., and Wigle, D., Herbicides and cancer. *JNCI* 84:1866-1874, 1992.
9. Bond, G.G., Bodner, K. M., and Cook, R. R., Phenoxy herbicides and cancer: Insufficient epidemiologic evidence for a causal relationship. *Fundam. Appl. Toxicol.* 12:172-188, 1989.
10. Axelson, O., Pesticides and cancer risks in agriculture. *Med. Oncol. Tumor Pharmacother.* 4:207-217, 1987.
11. Sharp, D. S. and Eskenazi, B., Delayed health hazards of pesticide exposure. *Annu. Rev. Public Health* 7:441-471, 1986.
12. Kay, K., Occupational cancer risks for pesticide workers. *Environ. Res.* 7:243-271, 1974.

13. Xue, S., Health effects of pesticides: A review of epidemiologic research from the perspective of developing nations. *Am. J. Ind. Med.* 12:269-279, 1987.

14. Burmeister, L. F., Cancer mortality in Iowa farmers, 1971-78. *JNCI* 66:461-464, 1981.

15a. Saftlas, A. F., Blair, A., Cantor, K. P., Hanrahan, L., and Anderson, H. A., Cancer and other causes of death among Wisconsin farmers. *Am. J. Ind. Med.* 11:119-129, 1987.

15b. Blair, A., Dosemeci, M., and Heineman, E. F., Cancer and other causes of death among male and female farmers from twenty-three states. *Am. J. Ind. Med.* 23:729-742, 1993.

16. Wiklund, K., Swedish agricultural workers. A group with decreased risk of cancer. *Cancer* 51:566-568, 1983.

17. Reif, J., Pearce, N., and Frazer, J., Cancer risks in New Zealand farmers. *Int. J. Epidemiol.* 18:768-774, 1989.

18. Ortega, G. L., Cancer en agricultores. Fondo de investigacion sanitaria. Madrid, Espana, 1991.

19. Stark, A. D., Chang, H., Fitzgerald, E. F., Riccardi, K., and Stone, R. R., A retrospective cohort study of mortality among New York state Farm Bureau members. *Arch. Environ. Health* 42:204-212, 1987.

20. Brownson, R. C., Reif, J. S., Chang, J. C., and Davis, J. R., Cancer risks among Missouri farmers. *Cancer* 64:2381-2386, 1989.

21. Delzell, E., Grufferman, S., Mortality among white and nonwhite farmers in North Carolina, 1976-1978. *Am. J. Epidemiol.* 121:391-402, 1985.

22. Gallagher, R. P., Threlfall, W. J., Jeffries, E., Band, P. R., Spinelli, J., and Coldman, A. J. Cancer and aplastic anemia in British Columbia farmers. *JNCI* 72:1311-1315, 1984.

23. Rafnsson, V. and Gunnarsdottir, H., Mortality among farmers in Iceland. *Int. J. Epidemiol.* 18:146-151, 1989.

24. Carlson, M. L. and Petersen, G.R., Mortality of California agricultural workers. *J. Occup. Med.* 20:30-32, 1978.

25. Almas, R. and Odegard, J., Morbidity among self-employed farmers in Norway. *Scand. J. Soc. Med.* 13:169-172, 1985.

26. Blair, A., Grauman, D. J., Lubin, J. H., and Fraumeni, J. F. Jr., Lung cancer and other causes of death among licensed pesticide applicators. *JNCI* 71:31-37, 1983.

27. Barthel, E., Increased risk of lung cancer in pesticide-exposed male agricultural workers. *J. Toxicol. Environ. Health* 8:1027-1040, 1981.

28. MacMahon, B., Monson, R. R., Wang, H. H., and Zheng, T. A., second follow-up of mortality in a cohort of pesticide applicators. *J. Occup. Med.* 30:429-43, 1988.

29. Wicklund, K. G., Daling, J. R., Allard, J., and Weis, N. S., Respiratory cancer among orchardists in Washington State, 1968 to 1980. *J. Occup. Med.* 30: 561-564, 1988.

30. Corrao, G., Calleri, M., Carle, F., Russo, R., Bosia, S., Piccion, I. P., Cancer risk in a cohort of licensed pesticide applicators. *Scand. J. Work Environ. Health* 15:203-209, 1989.

31. Alavanja, M. C. R., Blair, A., and Masters, M. N., Cancer mortality in the U.S. flour industry. *JNCI* 82:840-848, 1990.

32. Alberghini, V., Luberto, F., Gobba, F., Morelli, C., Gori, E., and Tomesani, N., Mortality among male farmers licensed to use pesticides. *Med. Lav.* 82:18-24, 1991.

33. Sathiakumar, N., Delzell, E., Austin, H., and Cole, P., A follow-up study of agricultural chemical production workers. *Am. J. Ind. Med.* 21:321-330, 1992.

34. Cantor, K. P. and Booze, C. F. Jr., Mortality among aerial pesticide applicators and flight instructors: A reprint. *Arch. Environ. Health* 46:110-116, 1991.

35. Hardell, L. and Sandstrom, A., Case-control study: Soft-tissue sarcomas and exposure to phenoxyacetic acids or chlorophenols. *Br. J. Cancer* 39:711-717, 1979.

36. Hardell, L., Eriksson, M,. Lenner, P., and Lundgren, E., Malignant lymphoma and exposure to chemicals, especially organic solvents, chlorophenols and phenoxy acids: A case-control study. *Br. J. Cancer* 43:169-176, 1981.

37. Donna, A., Betta, P., Robutti, F., Grosignani, P., Berrino, F., and Bellingeri, D., Ovarian mesothelial tumors and herbicides: A case-control study. *Carcinogenesis* 5:941-942, 1984.

38. Smith, A. H., Pearce, N. E., Fisher, D. O., Giles, H. J., Teague, C. A., and Howard, J, K., Soft tissue sarcoma and exposure to phenoxyherbicides and chlophenols in New Zealand. *JNCI* 73:1111-1117, 1984.

39. Eriksson, M., Hardell, L., Berg, N. O., Moller, T., and Axelson, O., Soft-tissue sarcomas and exposure to chemical substances: A case-referent study. *Br. J. Ind. Med.* 38:27-33, 1981.

40. Hardell, L., Johansson, B., and Axelson, O., Epidemiological study of nasal and nasopharyngeal cancer and their relation to phenoxy acid or chlorophenol exposure. *Am. J. Ind. Med.* 3:247-257, 1982.

41. Lynge, E., A follow-up study of cancer incidence among workers in manufacture of phenoxy herbicides in Denmark. *Br. J. Cancer* 52:259-270, 1985.

42. Pearce, N. E., Smith, A. H., Howard, J. K., Sheppard, R. A., Giles, H. J., and Teague, C. A., Non-Hodgkin's lymphoma and exposure to phenoxyherbicides, chlorophenols, fencing work, and meat works employment: A case-control study. *Br. J. Ind. Med.* 43:75-83, 1986.

43. Vineis, P., Terracini, B., Ciccone, G., Cignetti, A., Colombo, E., Donna, A., Maffi, L., Pisa, R., Ricci, P., Zanini, E., and Comba, P., Phenoxy herbicides and soft-tissue sarcomas in female rice weeders: A population-based case-referent study. *Scand. J. Work Environ. Health* 13:9-17, 1986.

44. Woods, J. S., Polissar, L., Severson, R. K., Heuser, L. S., and Kulander, B. G., Soft tissue sarcoma and non-Hodgkin's lymphoma in relation to phenoxyherbicide and chlorinated phenol exposure in western Washington. *JNCI* 78:899-910, 1987.

45. Hardell, L. and Eriksson, M., The association between soft-tissue sarcomas and exposure to phenoxyacetic acids: A new case-referent study. *Cancer* 62:652-656, 1988.

46. Donna, A., Crosignani, P., Robutte, F., Betta, P. G., Bocca, R., Mariani, N., Ferrario, F., Fissi, R., and Berrino, F., Triazine herbicides and ovarian epithelial neoplasms. *Scand. J. Work Environ. Health* 15:47-53, 1989.

47. Eriksson, M., Hardell, L., and Adami, H., Exposure to dioxins as a risk factor for soft-tissue sarcoma: A population-based case-control study. *JNCI* 82:486-490, 1990.

48. Wingren, G., Fredrikson, M., Brage, H. N., Nordenskjold, B., and Axelson, O., Soft tissue sarcoma and occupational exposures. *Cancer* 66:806-811, 1990.

49a. Wigle, D. T., Semenciw, R. M., Wilkins, K., Riedel, D., Ritter, L., Morrison, H. I., and Mao, Y., Mortality study of Canadian farm operators: Non-Hodgkin's lymphoma mortality and agricultural practices in Saskatchewan. *JNCI* 82:575-582, 1990.

49b. Morrison, H., Savitz, D., Semenciw, R., Hulka, B., Mao, Y., Morison, D., and Wigle, D., Farming and prostate cancer mortality. *Am. J. Epidemiol.* 137:270-280, 1993.

50. LaVecchia, C., Negri, E., D'Avanzo, B., and Franceschi, S., Occupation and lymphoid neoplasms. *Br. J. Cancer* 60:383-388, 1989.

51. Coggon, D., Pannett, B., and Winter, P., Mortality and incidence of cancer at four factories making phenoxy herbicides. *Br. J. Ind. Med.* 48:173-178, 1991.

52. Vineis, P., Faggiono, F., Tedeschi, M., and Ciccone, G., Incidence rates of lymphomas and soft-tissue sarcomas and environmental measurements of phenoxy herbicides. *JNCI* 83:362-363, 1991.

53. Green, L. M., A cohort mortality study of forestry workers exposed to phenoxy acid herbicides. *Br. J. Ind. Med.* 48:234-238, 1991.

54. Persson, B., Dahlander, A., Freriksson, M., Brage, H. N., Ohlson, C. G., Axelson, O., Malignant lymphomas and occupational exposures. Br. J. Ind. Med. 46:516-520, 1989.

55. Boffetta, P., Stellman, S. D., and Garfinkel, L., A case-control study of multiple myeloma nested in the American Cancer Society prospective study. *Int. J. Cancer* 43:554-559, 1989.

56. Cantor, K. P. and Blair, A., Farming and mortality from multiple myeloma: A case-control study with the use of death certificates. *JNCI* 72:251-255, 1984.

57. Morris, P. D., Koepsell, T. D., Daling, J. R., Taylor, J. W., Lyon, J. L., Swanson, G. M., Child, M., and Weiss, N. S., Toxic substance exposure and multiple myeloma: A case-control study. *JNCI* 76:987-994, 1986.

58. Flodin, U., Fredriksson, M., and Persson, B., Multiple myeloma and engine exhausts, fresh wood and creosote: A case-referent study. *Am. J. Ind. Med.* 12:519-529, 1987.

59. Williams, A. R., Weiss, N. S., Koepsell, T. D., Lyon, J. L., and Swanson, G. M., Infectious and noninfectious exposures in the etiology of light chain myeloma: A case-control study. *Cancer Res.* 49:4038-4041, 1989.

60. Ribbens, P. H., Mortality study of industrial workers exposed to aldrin, dieldrin and endrin. *Int. Arch. Occup. Environ. Health* 56:75-79, 1985.

61. Hearn, S., Ott, M. G., Kolesar, R. C., and Cook, R. R., Mortality experience of employees with occupational exposure to DBCP. *Arch. Environ. Health* 39:49-55, 1984.

62. Wong, O., Brocker, W., Davis, H. V., and Nagle, G. S., Mortality of workers potentially exposed to organic and inorganic brominated chemicals, DBCP, TRIS, PBB, and DDT. *Br. J. Ind. Med.* 41:15-24, 1984.

63. Ditraglia, D., Brown, D. P., Namekata, T., and Iverson, N., Mortality study of workers employed at organochlorine pesticide manufacturing plants. *Scand. J. Work Environ. Health* 7:7 (Suppl 4):140-146, 1981.

64. Ott, M. G., Scharnweber, H. C., and Langner, R. R., Mortality experience of 161 employees exposed to ethylene dibromide in two production units. *Br. J. Ind. Med.* 37:163-168, 1980.

65. Wang, H. H. and MacMahon, B., Mortality of workers employed in the manufacture of chlordane and heptachlor. *J. Occup. Med.* 21:745-748, 1979.

66. Mabuchi, K., Lilienfeld, A. M., and Snell, L. M., Cancer and occupational exposure to arsenic: A study of pesticide workers. *Prev. Med.* 9:51-77, 1980.

67. Austin, H., Keil, J. E., and Cole, P., A prospective follow-up study of cancer mortality in relation to serum DDT. *Am. J. Public Health* 79:43-46, 1989.

68. Saracci, R., Kogevinas, M., Bertazzi, P., De Mesquita, B. H. B., Coggon, D., Green, L. M., Kauppinen, T., L'Abbe, K. A., Littorin, M., Lynge, E., Mathews, J. D., Neuberger, M., Osman, J., Pearce, N., and Winkelmann, R., Cancer mortality in workers exposed to chlorphenoxy herbicides and chlorophenols. *Lancet* 338:1027-1032, 1991.

69. Fingerhut, M. A., Halperin, W. E., Marlow, D. A., Piacitelli, L. A., Honchar, P. A., Seeney, M. H., Greife, A. L., Dill, P. A., Steenland, K., and Suruda, A. J., Cancer mortality in workers exposed to 2,3,7,8-tetrachlorodibenzo-p-dioxin. *N. Engl. J. Med.* 324:212-218, 1991.

70. Cantor, K. P., Blair, A., Everett, G., Gibson, R., Burmeister, L. F., Brown, L. M., Schuman, L., and Dick, F. R., Pesticides and other agricultural risk factors for non-Hodgkin's lymphoma among men in Iowa and Minnesota. *Cancer Res.* 52:2447-2455, 1992.

71. Riihimaki, V., Asp, S., and Hernberg, S., Mortality of 2,4-dichlorophenoxyacetic acid and 2,4,5-trichlorophenocyacetic acid herbicide applicators in Finland. *Scand. J. Work Environ. Health* 8:37-42, 1982.

72. Axelson, O., Sundell, L., Andersson, K., Edling, C., Hogstedt, C., and Kling, H., Herbicide exposure and tumor mortality. *Scand. J. Work Environ. Health* 6:73-79, 1980.

73. Bond, G. G., Wetterstroem, N. H., Roush, G. J., McLaren, E. A., Lipps, T. E., and Cook, R. R., Cause specific mortality among employees engaged in the manufacture, formulation, or packaging of 2,4-dichlorophenoxyacetic acid and related salts. *Br. J. Ind. Med.* 45:98-105, 1988.

74. Hoar, S. K., Blair, A., Holmes, F. F., Boysen, C. D., Robel, R. J., Hoover, R., and Fraumeni, J. F. Jr., Agricultural herbicide use and risk of lymphoma and soft-tissue sarcoma. *J. Am. Med. Assoc.* 256:1141-1147, 1986.

75. Coggon, D., Pannett, B., Winter, P. D., Acheson, E. D., and Bonsall, J., Mortality of workers exposed to 2 methyl-4 chlorophenoxyacetic acid. *Scand. J. Work Environ. Health* 12:448-454, 1986.

76. Ott, M. G., Holder, B. B., and Olson, R. D., A mortality analysis of employees engaged in the manufacture of 2,4,5-trichlorophenoxyacetic acid. *J. Occup. Med.* 22:47-50, 1980.

77a. Zahm, S. H., Weisenburger, D. D., Babbitt, P. A., Saal, R. C., Vaught, J. B., Cantor, K. P., and Blair, A., A case-control study of non-Hodgkin's lymphoma and the herbicide 2,4-dichlorophenoxyacetic acid (2,4-D) in eastern Nebraska. *Epidemiology* 1:349-356, 1990.

77b. Zahm, S. H, Weisenburger, D. D., Saal, R. C., Vaught, J. B., Babbitt, P. A., and Blair, A., Pesticide use, genetic susceptibility, and non-Hodgkin's lymphoma in women. In: McDuffie, H. H., Dosman, J. A., Semchuk, K., Olenchock, S., and Seuthilselvan, A. (Eds). *Agriculture Health and Safety: Workplace, Environment, and Sustainability.* CRC Press/Lewis Publishers, Boca Raton, FL., 1995, 127.

77c. Zahm, S. H., Blair, A., and Weisenburger, D. D., Sex differences in the risk of multiple myeloma associated with agriculture. *Br. J. Ind. Med.* 49:815-816, 1992.

78a. Brown, L. M., Blair, A., Gibson, R., Everett, G. D., Cantor, K. P., Schuman, L. M., Burmeister, L. F., Van Lier, S. F., and Dick, F., Pesticide exposures and other agricultural risk factors for leukemia among men in Iowa and Minnesota. *Cancer Res.* 50:6585-6591, 1990.

78b. Brown, L. M., Burmeister, L. F., Everett, G. D., and Blair, A., Pesticide exposures and multiple myeloma in Iowa men. *Cancer Causes Control* 4:153-156, 1993.

79. Zahm, S. H., Weisenburger, D. D., Cantor, K. P., Holmes, F. F., and Blair, A., Role of the herbicide atrazine in the development of non-Hodgkin's lymphoma. *Scand. J. Work Environ. Health* 19:108-114, 1993.

80. Garabrant, D. H., Held, J., Bangholz, B., Peters, J. M., and Mack, T. M., DDT and related compounds and risk of pancreatic cancer. *JNCI* 84:764-771, 1992.

81. Eriksson, M. and Karlsson, M., Occupational and other environmental factors and multiple myeloma: A population based case-control study. *Br. J. Ind. Med.* 49:95-103. 1992.

82a. Falck, F., Ricci, A., Wolff, M. S., Godbold, J., and Deckers, P., Pesticides and polychlorinated biphenyl residues in human breast lipids and their relation to breast cancer. *Arch. Environ. Health* 47:143-146, 1992.

82b. Wolf, M. S., Toniolo, P. G., Lee, E. W., Rivera, M., and Dubin, N., Blood levels of organochlorine residues and risk of breast cancer. *JNCI* 85:648-652, 1993.

83. International Agency for Research on Cancer. *IARC monographs on the evaluation of carcinogenic risks to humans. Overall evaluations of carcinogenicity: an updating of IARC Monographs Volumes 1 to 42.* Supplement 7. Lyon, France. 1987.

84. International Agency for Research on Cancer. *IARC monographs on the evaluation of carcinogenic risks to humans. Occupational exposures in insecticide application and some pesticides. Volume 53.* Lyon, France. 1991.

85. Garrett, N. E., Stack, H. F., and Waters, M. D., Evaluation of the genetic activity profiles of 65 pesticides. *Mutat. Res.* 168:301-325, 1986.

86. Moser, G. J., Meyer, S. A., and Smart, R. C., The chlorinated pesticide Mirex is a novel nonphorbol ester-type tumor promoter in mouse skin. *Cancer Res.* 52:631-636, 1992.

87. Kniewald, J., Mildner, P., and Kniewald, Z., Effects of s-triazine herbicides on hormone-receptor complex formation, 5 alpha-reductase and 3 alpha-hydroxysteriod dehydrogenase activity at the anterior pituitary level. *J. Steroid Biochem.* 11:833-838, 1979.

88. Vineis, P. and D'Amore, F., The role of occupational exposure and immunodeficiency in B-cell malignancies. *Epidemiology* 3:266-270, 1992.

89. Newcombe, D. S., Immune surveillance, organophosphorus exposure and lymphomagenesis. *Lancet* 339:539-541, 1992.

90. Wang, R. G. M., Franklin, C. A., Honeycutt, R. C., and Reinert, J. C. (Eds.). *Biological Monitoring for Pesticide Exposure.* Am. Chem. Soc. Series 382. American Chemical Society, Washington, D.C., 1989.

91. Blair, A., Zahm, S. H., Cantor, K. P., and Stewart, P. A., Estimating exposure to pesticides in epidemiologic studies of cancer. In: *Biological Monitoring for Pesticide Exposure*. Wang, R.G.M., Franklin, C.A., Honeycutt, R.C., and Reinert, J.C. (Eds.). Am. Chem. Soc. Series 382. American Chemical Society, Washington, D.C., 1989, pp. 38-46.

92. Blair, A. and Zahm, S. H., Patterns of pesticide use among farmers: Implications for epidemiologic research. *Epidemiology* 4:55-62, 1993.

93. Checkoway, H., Pearce, N. E., and Crawford-Brown, D. J., *Research Methods in Occupational Epidemiology*. Oxford University Press, New York, NY, 1989.

94. Copeland, K. T., Checkoway, H., McMichael, A. J., and Holbrook, R. H., Bias due to misclassification in the estimation of relative risk. *Am. J. Epidemiol.* 105:488-495, 1977.

95. Gold, E., Gordis, L., Tonascia, J., and Szklo, M., Risk factors for brain tumors in children. *Am. J. Epidemiol.* 109:309-319, 1979.

96. Wilkins, J. R. and Koutras, R. A., Paternal occupation and brain cancer in offspring: A mortality-based case-control study. *Am. J. Ind. Med.* 14:299-318, 1988.

97. Lowengart, R. A., Peters, J. M., Cicioni, C., Buckley, J., Bernstein, L., Preston-Martin, S., and Rappaport, E., Childhood leukemia and parents' occupational and home exposures. *JNCI* 79:39-46, 1987.

98. Shu, X. O., Gao, Y. T., Brinton, L. A., Linet, M. S., Tu, J. T., Zheng, W., and Fraumeni, J. F. Jr., A population-based case-control study of childhood leukemia in Shanghai. *Cancer* 62:635-644, 1988.

99. Buckley, J. D., Robison, L. L., Swotinsky, R., Garabrant, D. H., LeBeau, M., Manchester, P., Nesbit, M. E., Odom, L., Peters, J. M., Woods, W. G., and Hammond, G. D., Occupational exposures of parents of children with acute nonlymphocytic leukemia: A report from the Childrens Cancer Study Group. *Cancer Res.* 49:4030-4037, 1989.

100. Holly, E. A., Aston, D. A., Ahn, D. K., and Kristiansen, J. J., Ewing's bone sarcoma, paternal occupational exposure, and other factors. *Am. J. Epidemiol.* 135:122-129, 1992.

101. Sellers, T. A., Potter, J. D., Bailey-Wilson, J. E., Rich, S. S., Rothschild, H., and Elston, R. C., Lung cancer detection and prevention: Evidence for an interaction between smoking genetic predisposition. *Cancer Res.* 52 (Suppl):2694s-2697s, 1992.

102. Ooi, W. L., Elston, R. C., Chen, V. W., Bailey-Wilson, J. E., and Rothschild, H., Increased familial risk for lung cancer. *JNCI* 76:217-222, 1986.

OVERVIEW OF EVIDENCE AND RESEARCH NEEDS CONCERNING ELECTROMAGNETIC FIELDS AND HEALTH

7

David A. Savitz

CONTENTS

BASIS FOR CONCERN

Concern with potential adverse health effects from exposure to electro-magnetic fields comes largely from epidemiologic studies of power-frequency electric and magnetic fields. Laboratory studies of health effects of such fields have only begun recently, with most of the previous laboratory research oriented towards basic biophysics rather than toxicology. Thus, at the present time, mechanisms by which such exposures might cause disease are speculative with limited theoretical or empirical support. The pathways mediated by tissue heating or ionisation are clearly not applicable to the frequencies and intensities of concern, requiring a more subtle biophysical process to be involved to account for the epidemiologic indications of adverse health effects. Integration of laboratory and epidemiologic evidence is thus an important research goal in this area.

Independent of the level of evidence supporting adverse health effects, the prevalence of these exposures provides a compelling argument to conduct the needed research to address that possibility. Throughout the developed world, distribution and use of electricity results in exposure to power-frequency (50- or 60-Hz) electric and magnetic fields. Electric fields are the product of the voltage or electrical charge, and magnetic fields result from the movement of those charges or current. Wherever there are power lines, electrical appliances, and electrical wiring, some level of fields will be present. There are few agents of environmental health concern so ubiquitous that some detectable levels are certain to be present in every home or workplace. Higher frequency fields are also highly prevalent, ranging from sources such as video display terminals, televisions, cellular telephones, microwave transmitters, and radio and television transmitters.

Epidemiologic studies of power-frequency fields have focused on sources of prolonged exposure, which are more limited in variety. A dominant determinant of total magnetic field exposure is the ambient level in the residence, determined largely by the nearby electric power lines, grounding practices, and wiring in the home. Occupations that involve working in close proximity to energised equipment will produce elevated exposures to electric and magnetic fields, with a diverse array of occupations potentially exposed including power line workers, appliance repair persons, video display terminal users, and electric railway workers. A third source of exposure is electrical appliances, but only those that are used for prolonged periods such as electric blankets or underpads or certain types of bedside electric clocks. Transient exposures received from diverse sources such as hair dryers or electric razors produce brief elevations in exposure, but have not been the focus of epidemiologic study.

More than for most environmental health issues, this research has a direct impact on public policy, whether or not adverse health effects are ultimately identified. Societal decisions cannot be postponed until scientific consensus is reached, since installation of high-tension power lines, design of video display terminals, and home buyers' decisions about living near existing power lines make implicit or explicit assumptions about the importance of exposures to non-ionising radiation. Thus, in many countries, the economic and industrial interests of electric utility companies, government regulators, and the public force the issue to the forefront. It has been argued that this issue is a low priority relative to other health research (CIRRPC, 1993) given the tenuous biological support for harmful effects. However, when the expense of additional research is weighed against the tremendous societal costs of uncertainty (Florig, 1992), the need for information is well-justified whether we ultimately prove the presence or absence of harm from these fields.

EPIDEMIOLOGIC EVIDENCE

Cancer

Residential Exposures and Childhood Cancer

The most attention on electromagnetic fields and health has been focused on the possible association between residential exposure to power-frequency magnetic fields and childhood cancer. Studies have linked such exposures to leukaemia, brain cancer, and other cancers of children. Positive associations with leukaemia have been reported, with 1.5- to 3-fold risks found in the U.S. (Wertheimer & Leeper, 1979; Savitz et al., 1988; London et al., 1991), and most recently in Sweden (Feychting & Ahlbom, 1992). Positive associations have been reported with brain cancer as well (Wertheimer & Leeper, 1979; Savitz et al., 1988; Tomenius, 1986). Other studies have not supported an association (Fulton et al., 1980) and specific cancer types have not been associated with estimated exposure such as leukaemia in the study by Tomenius (1986) and brain cancer in the study by Feychting and Ahlbom (1992). The studies in which associations were not reported have generally been less persuasive than those finding associations, particularly the one by Fulton et al. (1980) in which control selection was biased relative to the chosen cases (Wertheimer & Leeper, 1980). Thus, this line of evidence is sufficiently credible and important to warrant more detailed discussion.

This research avenue was initiated by Wertheimer and Leeper (1979) with their report of a case-control study in Denver, Colorado. Residences of 344 children who had died of childhood cancer were compared to residences of an equal number of healthy children selected from birth certificate files. In the era of this study, childhood cancer was virtually always fatal. Exposure was characterised by a wiring configuration code, a method which was viewed rather skeptically initially.

The investigators reasoned that current flow along the power lines outside the home would influence magnetic fields inside the home. (Electric fields are effectively shielded by objects between the line and the home, including trees and building materials, whereas magnetic fields are unperturbed by such barriers.) Furthermore, they argued that the current flow along those lines could be inferred based on observable characteristics of the line including the location of transformers and service drops and the thickness of the line. In simple terms, lines carrying electricity to many homes are likely to have more current flowing on average than lines carrying electricity to few or no homes. With an estimate of current flow based on the line configurations and an estimate of the distance from the lines to the home, residences were classified into high and low wire codes for analysis. A selected set of measurements of magnetic fields in the vicinity of homes was obtained which generally supported the approach to classification: homes expected to have higher magnetic fields showed higher measurements.

The results of the study were notably and consistently positive. Although not presented in the article, calculated odds ratios associating childhood cancer with high wire code configurations were in the range of two to three based on residence at birth or diagnosis. This magnitude of increased risk was quite consistent across cancer types, gender, and child's age. The available confounders based on birth certificate (limited to demographic factors such as mother's age, education, and race) did not account for the observed association.

It is convenient to use the original Wertheimer and Leeper (1979) report as the starting point to describe the evolution of the research. As might be expected in response to a rather novel hypothesis and observation, criticism was abundant.

One of the strongest arguments that their study may have been biased towards a positive association was that the investigators, who made their own wire code assignments with knowledge of the case or control status of the occupant, were subtly biased. Subsequent studies in Denver (Savitz et al., 1988), Los Angeles (London et al., 1991), and Stockholm (Tomenius, 1986; Feychting & Ahlbom, 1992) have eliminated that potential source of bias by having the coding or measurements done by a technician who is unaware of the health status of the occupants. Thus, positive findings in those later studies argue against bias by the data collectors as the source of spurious positive results.

Potential confounders have been a concern in a number of studies. Several features of wire configurations and childhood cancer make this a difficult issue to resolve. There is an impression that higher current configurations are typical of poorer sections of the community with less desirable housing. Such a perception might argue that it is some other correlate of unfavourable economic circumstances that results in an increased risk for childhood cancer such as poorer diet, crowding, or other urban pollutants. This issue has not been entirely resolved, in that "wire configuration code" may be a proxy for other unidentified exposures as well as reflecting average magnetic fields. The failure of studies that have included actual measurements of magnetic fields in the home to yield stronger associations with cancer than those identified by wire configuration codes (see below) adds to the concern with confounding.

The lack of knowledge regarding causes of childhood cancers makes it difficult to speculate about specific confounders. In studies that have obtained interviews for a wide variety of potential risk factors (Savitz et al., 1988; London et al., 1991), there has been little or no indication that confounding was present. Also, the lack of specificity for cancer types in Wertheimer and Leeper's (1979) study raised some suspicions of possible bias, although in the absence of an established pattern of risk factors for childhood cancer, it is unclear whether or not to expect shared risk factors across cancer types.

Several studies have also considered exposure through use of electric appliances, particularly electric blankets and heated water beds. Savitz et al. (1990) reported a modest positive association between electric blanket use and childhood cancer both for in utero and postnatal exposure. London et al. (1991) found a number of appliances to be linked to childhood leukaemia, particularly black and white televisions and hair dryers.

Residential Exposures and Adult Cancer

Much less attention has been given to adult as compared to childhood cancers. Wertheimer and Leeper (1982) also pursued this avenue, with a study suggesting positive associations for many cancer types. Subsequent studies have generally not supported associations between residential exposure and adult leukaemia (Severson et al., 1988; Coleman et al., 1989; McDowall, 1986), although few cancer sites have been studied adequately. The methods have been similar to those employed in childhood cancer studies, relying on wire configuration codes and magnetic field measurements to classify exposure. For adults, the potential for confounding by indices of social class and occupation is more apparent. In addition, studies have begun to address appliance exposure, particularly electric blankets, in relation to testicular cancer (Verreault et al., 1990), adult leukaemia (Preston-Martin et al., 1988), and female breast cancer (Vena et al., 1991), with largely negative results.

Occupational Exposure and Adult Cancer

Virtually independent of the research on residential exposure to magnetic fields, many reports of cancer among electrical workers have been published. The first of these was a letter to the editor of the *New England Journal of Medicine* by Milham (1982), which demonstrated an increase in proportionate mortality from leukaemia, particularly acute myeloid leukaemia, among a group of electrical workers. The constitution of the group was defined based on intuition about workers who spent time in proximity to electrical equipment, including power and telephone linemen, electricians, streetcar and subway motormen, radio and radar operators, etc.

The literature on leukaemia and brain cancer among various subgroups of electrical workers now consists of around 50 studies including many letters to the editor but a number of complete manuscripts as well. Older (Savitz & Calle, 1987) and more recent reviews (Theriault, 1991; Savitz & Ahlbom, 1993) document a small increase in leukaemia incidence and mortality among electrical workers in the aggregate. The risk ratio is on the order of 1.2 overall, but tends to be somewhat higher (around 1.4) for acute myeloid leukaemia. Individual groups of electrical workers show higher or lower risk ratios, with linemen most consistently found to be at increased risk (Savitz & Ahlbom, 1993). Some of the case-control studies, in particular, have found marked increases in risk among electrical workers (Stern et al., 1986; Bastuji-Garin et al., 1990).

Brain cancer studies have generally been more ambitious in attempting to categorise the certainty with which specific jobs produce increases in exposure to electromagnetic fields, starting with the initial report by Lin et al. (1985). Although not perfectly consistent, there is replicated evidence of increased risks of brain tumours among electrical workers (Theriault, 1991; Savitz & Ahlbom, 1993), with some individual studies showing marked increases in risk (Speers et al., 1988; Mack et al., 1991).

Recently, a series of reports have considered male breast cancer among electrical workers (reviewed by Savitz & Ahlbom, 1993). Although this cancer is extremely rare, the hypothesised link to pineal melatonin production gives this cancer site particular interest. Other cancers have been considered only sporadically in relation to electrical work, including lymphomas and melanoma, with limited support for an association.

Methodological concerns with this literature are focused principally on exposure assessment. With few exceptions, this is actually the epidemiology of "job titles" with the link to exposures assumed rather than proven. As studies begin to examine actual workplace exposures associated with electrical occupations, some jobs are found to have such exposures and others are not (Bowman et al., 1988). For example, electricians who work in construction do not typically spend time near energised equipment, whereas those who work in electrical substations or power plants do have elevated exposures. To the extent that job title is an imperfect surrogate for elevated exposure and exposure actually increases the risk of cancer, these studies will yield diluted measures of association with cancer since their design ensures that misclassification will be similar for cancer cases and non-cases. The possibility that reported risk ratios of 1.5 or 2.0 may be accounted for by markedly larger risk ratios among the truly exposed subset of workers is a major motivation to pursue this line of research.

Other methodological concerns have also been raised. Proportionate mortality or incidence is known to be vulnerable to exaggeration as an indicator of the risk ratio (Wong & Decoufle, 1982) when other causes of death or disease are rarer among the study population, although Coleman and Beral (1988) found electrical workers in the aggregate in England to have total mortality similar to the general population. Other workplace exposures to chemicals might act as confounders, since solvents and other potentially carcinogenic chemicals are used in some electrical occupations though not in great quantity. Finally, a concern has been raised that the published positive results represent some form of publication bias, in which those whose data bases do not support the association do not bother to publish their findings (National Radiation Protection Board, 1992). This is possible but highly unlikely since most of the data bases suitable for such analyses have been exploited and following the initial positive reports, negative findings are of real interest both to investigators and journal editors.

Reproductive Effects

The epidemiologic literature on reproductive health consequences of electromagnetic fields exposure has concentrated on the use of video display terminals (VDTs). VDTs produce modest elevations in power frequency fields, but markedly greater elevations in fields in the kilohertz range. Like many other workplace exposures, it is difficult to separate any effect of fields from VDTs from correlated aspects of the workplace, including psychological stress, sedentary nature of the job, and ergonomic considerations.

Referring to recent reviews for more details (Hatch, 1992; Shaw & Croen, 1993), the literature overall provides little support for an association with increased risk of miscarriage, fetal growth retardation, preterm delivery, or birth defects among offspring. Individual studies have suggested modest increases in miscarriage for women working 20 or more hours per week with VDTs, but the more recent and sophisticated studies (Schnorr et al., 1991; Roman et al., 1992) do not support that contention. Although VDT work in general may not confer an increase in risk, actual exposure to non-ionising radiation associated with VDT work varies markedly depending on the placement of nearby VDTs and the model and age of the machine. Thus, the question of whether *exposures* from VDTs influence reproductive health has not been answered conclusively.

Other sources of power frequency fields have received relatively little attention. Electric blankets and related sources have been considered in studies of miscarriage and fetal growth (Wertheimer & Leeper, 1989), with suggestive seasonal effects related to appliance use, and a completely negative report on birth defects (Dlugosz et al., 1992). Given the concern with VDTs, additional research on residential exposures, appliance use, and even more pronounced occupational exposures to pregnant women in relation to reproductive outcomes is warranted.

Behavioural/Neuropsychological Effects

If one were selecting an epidemiologic research avenue based on the laboratory research and hypothesised mechanisms, the most promising would likely be behavioural and neuropsychological effects. Direct influences on nervous system function from electromagnetic field exposures and indirect effects mediated by hormones are applicable. In fact, some of the earliest reports of adverse health effects were from switchyard workers in Russia in the 1960s (Danilin et al., 1969).

Research on cognitive and behavioural responses to fields has improved and grown markedly, culminating in human experiments largely demonstrating an absence of effects in volunteers (Stollery, 1986). Clinical consequences of such exposures have received less attention, with some reports of suicide and depression in relation to residential exposure to magnetic fields (Reichmanis et al., 1979; Perry et al., 1989) but without even an attempt at replication. In the aggregate, the epidemiologic literature on cognitive and behavioural consequences of electric and magnetic field exposure is disjointed and weak, providing little guidance to future investigators (Paneth, 1993).

RESEARCH NEEDS

Mechanism of Action: Laboratory Studies Suggested by Epidemiology

A persistent concern in attempting to measure residential or occupational magnetic field exposures is uncertainty regarding the biologically relevant exposure parameter. In the face of ignorance, researchers gravitate towards some average or time-integrated total exposure; yet under different biophysical theories, peaks, time above thresholds, fluctuation in field intensity, or even windows of intensity may be more important. Both wire configuration codes and measured average fields in homes are likely to be correlated with such alternative exposure metrics but not perfectly. Similarly, the job title of "electrical worker" generally conveys some information about exposure, but there is very little work other than in the electric utility industry that empirically examines what exposure metrics are reflected by specific jobs.

Epidemiologists can and should attempt to specify exposure characteristics more precisely and develop hypotheses regarding what exposure circumstances are likely to influence health. However, there is a real limitation in our inability to manipulate exposure, such that we can only evaluate circumstances that are naturally occurring. Clearly, laboratory studies of carcinogenesis, in particular, are needed to address the question of the proper exposure metric. If a relevant experimental system could be established, then the questions about temporal patterns of exposure, thresholds or ceiling of effect, joint effects of electric and magnetic fields, etc., could be addressed with far greater precision than could ever be obtained through epidemiology.

The challenges to addressing health effects of electromagnetic fields in the laboratory should not be underestimated. The logistics of setting up exposure apparatus requires extensive engineering collaboration. The biological processes likely to be affected are unclear given that the agent is known not to be mutagenic but would have to influence cancer risk through some other processes. There is a lack of a clear exposure metric based on epidemiologic observations, such that the high-dose counterpart of "wire code" or "electrical occupation" is not clearly defined. Nonetheless, toxicological efforts to parallel those in epidemiology are badly needed.

Control Selection in Residential Studies of Cancer

One of the principal challenges raised regarding the reported associations between wiring configuration codes and childhood cancers is the ever-present possibility that the controls do not accurately reflect the composition of the study base. Poole and Trichopoulos (1991) particularly have put forth the possibility that through telephone-based selection and high rates of nonresponse, the resulting control group may underrepresent high wire code homes and thus spuriously create an association between wire code and childhood cancer. Studies conducted in settings that are more amenable to selecting representative controls (e.g., Scandinavia) are attractive sites for further research.

Exposure and Disease Assessment

Methods for Exposure Classification in Epidemiologic Studies

Exposure assessment was and remains the greatest challenge and source of concern. Initially, questions were raised about whether the indirect information about wiring near the home would have any relationship to actual magnetic fields in the home. There is rather convincing evidence from studies in Denver (Wertheimer & Leeper, 1979; Savitz et al., 1988), Seattle (Kaune et al., 1987), and Los Angeles (London et al., 1991) that wire codes are predictive of measured fields, but only moderately so.

A major controversy at present comes from the observation that studies that have included both measured fields and wire codes have tended to find stronger support for an association of wire codes with childhood cancer than for measured magnetic fields. It must be remembered that both of these are merely proxies for the historical exposures of interest. Although wire codes are indirect markers of magnetic fields, they are historically stable. In contrast, measured fields are affected by short-term influences such as the electric power consumption at the time of measurement and changes in in-home wiring that limit their relevance to historical levels. Thus, wire codes may better reflect the true relationship of exposure with disease. On the other hand, it might be speculated that the stronger relationship of wire codes with cancer reflects confounding, with measured fields giving a less biased measure of association. Data on the historical fields of interest are needed to distinguish between these explanations.

In the occupational setting, a more familiar challenge applies. Since work histories (job titles, periods of employment) remain the principal index to historical exposures, improved job-exposure matrices linking those jobs to exposure are needed. Past efforts at linkage have been largely intuitive, the most extreme being the one in which an electrical worker job title is equated with exposure. Not only should electrical jobs be separated from one another, with exposure probabilities, levels, and field types differing across work settings, but even within those jobs, the tasks and time periods over which they were held might well refine classification further.

Finally, the structure of past residential and occupational studies of electric and magnetic fields has precluded considering the two exposure sources in the same study. The available data suggest that sizeable contributions to total time-integrated exposure come from both sources (Deadman et al., 1988), such that exposure classification would be reduced if they could be considered in combination.

Cancer Outcomes

The selection and categorisation of cancer outcomes has been driven largely by prior epidemiologic observations rather than an hypothesised biological process. Epidemiologic leads will continue to provide the most important guidance, but based on the study of other environmental carcinogens, some refinements can be suggested. Certainly, the histologic types of leukaemia and brain cancer should be separated where possible for analysis. Further refinements based on cytogenetic classifications could yield even more etiologically homogeneous entities. Ultimately, such subgroups may show similar results and thus warrant reaggregation, but unless the finer classifications are considered, the degree of specificity in the associations will not be known.

Pineal Melatonin Hypothesis

The hypothesised influence of electric and magnetic fields on pineal function (Stevens et al., 1992) points specifically to cancer types and other health outcomes that have received little previous epidemiologic attention. The most compelling from the public health perspective are female breast cancer and depression, both tied to hormones influenced by pineal melatonin. The epidemiology alone on these outcomes would provide only modest encouragement for additional study, but the epidemiology combined with clearly postulated pathways lends much greater support for such studies.

Vulnerable Groups

Little is presently known about which groups, if any, are particularly vulnerable to electromagnetic fields. When the aggregate effect is unclear based on the epidemiology and biological mechanisms are poorly understood, it is difficult to even speculate about which groups would be most likely to suffer adverse effects of exposure.

Children have been of much greater interest than adults in studies of residential exposure and cancer; thus it might be postulated that children are more vulnerable, are more amenable to study since they spend more time in the home, or simply have been arbitrarily selected for greater attention than adults.

Since electromagnetic fields are clearly not mutagenic, it might be postulated that populations who have mutations that predispose them to cancer would be especially vulnerable to electric and magnetic fields. However, except for rare genetic syndromes, there are no practical means of identifying vulnerable individuals.

Finally, given the diversity of field frequencies, amplitudes, and temporal patterns found in the human environment, it is quite likely that if any have adverse health effects, the specific forms of exposure will vary in their potency. Although not a vulnerable population due to their inherent characteristics, persons experiencing the most potent form of the agent will be more vulnerable to any adverse health consequences.

HYPOTHESES TO BE TESTED

There are a wide variety of research avenues that could be pursued based on epidemiologic, laboratory, or theoretical considerations. However, in my opinion, only a few epidemiologic research avenues are compelling and certain to be informative regardless of whether positive or negative findings are obtained:

1. The link between residential exposure to magnetic fields, as reflected in part by wiring configuration codes, and childhood cancer needs to be confirmed or refuted by conducting studies that address specific methodological deficiencies in past studies. By specifying and testing such alternative explanations, the tenability of a causal association will either be strengthened if such biases are shown not to account for the results or challenged if such biases are found to operate.
2. Occupational influences on adult leukaemia and brain cancer need to be pursued, principally by defining exposure more precisely and assessing whether the strength of association is actually related to exposures to electric and magnetic fields or whether some other correlate of those job titles is actually accounting for the associations reported in past studies.
3. The consequences of exposure predicted from laboratory research warrant epidemiologic testing, specifically those based on alterations on pineal melatonin. Most compelling would be clinical or laboratory studies that bear upon the possibility of breast cancer and clinical depression.

REFERENCES

Bastuji-Garin, S., Richardson, S., and Zittoun, R., Acute leukaemia in workers exposed to electromagnetic fields, *Eur. J. Cancer*, 26, 1119, 1990.

Bowman, J. D., Garabrant, D. H., Sobel, E., and Peters, J. M., Exposures to extremely low frequency (ELF) electromagnetic fields in occupations with elevated leukemia rates, *Appl. Ind. Hyg.*, 3, 189, 1988.

CIRRPC (The Committee on Interagency Radiation Research and Policy Coordination), Health effects of low frequency electric and magnetic fields, (Executive Summary), *Environ. Sci. Technol.*, 27, 42, 1993.

Coleman, M. and Beral, V., A review of epidemiological studies of the health effects of living near or working with electricity generation and transmission equipment, *Int. J. Epidemiol.*, 17, 1, 1988.

Coleman, M. P., Bell, C. M., Taylor, H. L., and Primic-Zakelj, M., Leukaemia and residence near electricity transmission equipment: a case- control study, *Br. J. Cancer*, 60, 793, 1989.

Danilin, V. A., Voronin, A. K., and Madorskii, V. A., The state of health of personnel working in high voltage electric fields, *Gig. Tr. Prof. Zabol.*, 13, 51, 1969.

Deadman, J. E., Camus, M., Armstrong, B. G., Heroux, P., Cyr, D., Plante, M., and Theriault, G., Occupational and residential 60-Hz electromagnetic fields and high-frequency transients: exposure assessment using a new dosimeter, *Am. Ind. Hyg. Assoc. J.*, 49, 409, 1988.

Dlugosz, L., Vena, J., Byers, T., Sever, L., Bracken, M., and Marshall, E., Congenital defects and electric bed heating in New York State: a register- based case-control study, *Am. J. Epidemiol.*, 135, 1000, 1992.

Feychting, M. and Ahlbom, A., Magnetic fields and cancer in people residing near Swedish high voltage power lines, Report from the Karolinska Institute, Stockholm, 1992.

Florig, H. K., Containing the costs of the EMF problem, *Science*, 257, 468, 1992.

Fulton, J. P., Cobb, S., Preble, L., Leone, L., and Forman, E., Electrical wiring configurations and childhood leukemia in Rhode Island, *Am. J. Epidemiol.*, 111, 292, 1980.

Hatch, M., The epidemiology of electric and magnetic field exposures in the power frequency range and reproductive outcomes, *Paediatr. Perinatal Epidemiol.*, 6, 198, 1992.

Kaune, W. T., Stevens, R. G., Callahan, N. J., Severson, R. K., and Thomas, D.B., Residential magnetic and electric fields, *Bioelectromagnetics*, 8, 315, 1987.

Lin, R. S., Dischinger, P. C., Conde, J., and Farrell, K. P., Occupational exposure to electromagnetic fields and the occurrence of brain tumors, *J. Occup. Med.*, 27, 413, 1985.

London, S. J., Thomas, D. C., Bowman, J. D., Sobel, E., Cheng, T.-C., and Peters, J.M., Exposure to residential electric and magnetic fields and risk of childhood leukemia, *Am. J. Epidemiol.*, 134, 923, 1991.

Mack, W., Preston-Martin, S., and Peters, J. M., Astrocytoma risk related to job exposure to electric and magnetic fields, *Bioelectromagnetics*, 12, 57, 1991.

McDowall, M. E., Mortality of persons resident in the vicinity of electricity transmission facilities, *Br. J. Cancer*, 53, 271, 1986.

Milham S., Mortality from leukemia in workers exposed to electrical and magnetic fields (Letter), *N. Engl. J. Med.*, 307, 249, 1982.

Myers, A., Clayden, A. D., Cartwright, R. A., and Cartwright, S. C., Childhood cancer and overhead powerlines. A case/control study, *Br. J. Cancer* 62, 1008, 1990.

National Radiation Protection Board, Electromagnetic Fields and the Risk of Cancer. National Radiation Protection Board, Chilton, England, 1992.

Paneth, N., Neurobehavioral effects of power frequency electromagnetic fields, *Environ. Health Perspect.*, 1993, 101 (Suppl. 4), 101, 1993.

Perry, S., Pearl, L., and Binns, R., Power frequency magnetic field: depressive illness and myocardial infarction, *Public Health*, 103, 177, 1989.

Poole, C., and Trichopoulos, D., Extremely low-frequency electric and magnetic fields and cancer, *Cancer Causes Control*, 2, 267, 1991.

Preston-Martin, S., Peters, J. M., Yu, M. C., Garabrant, D. H., and Bowman, J. D., Myelogenous leukemia and electric blanket use, *Bioelectromagnetics*, 9, 207, 1988.

Reichmanis, M., Perry, F. S., Marino, A. A., and Becker, R. O., Relation between suicide and the electromagnetic field of overhead power lines. *Physiol. Chem. Phys.*, 11, 395, 1979.

Roman, E., Beral, V., Pelerin, M., and Hermon, C., Spontaneous abortion and work with visual display units, *Br. J. Ind. Med.*, 49, 507, 1992.

Savitz, D. A. and Ahlbom, A., Epidemiologic evidence on cancer in relation to residential and occupational exposures, in *Biologic Effects of Electric and Magnetic Fields*, Carpenter, D. O., Ed., Academic Press, New York, 1993.

Savitz, D. A. and Calle, E. E., Leukemia and occupational exposure to electromagnetic fields: review of epidemiological surveys, *J. Occup. Med.*, 29, 47, 1987.

Savitz, D. A., Wachtel, H., Barnes, F. A., John, E. M., and Tvrdik, J. G., Case-control study of childhood cancer and exposure to 60-hertz electric and magnetic fields, *Am. J. Epidemiol.*, 128, 21, 1988.

Savitz, D. A., John, E. M., and Kleckner, R. C., Magnetic field exposure from electric appliances and childhood cancer, *Am. J. Epidemiol.*, 131, 763, 1990.

Schnorr, T. M., Grajewski, B. A., Hornung, R. W., Thun, M. J., Egeland, G. M., Murray, W. E., Conover, D. L., and Halperin, W. E., Video display terminals and the risk of spontaneous abortion, *N. Engl. J. Med.*, 324, 727, 1991.

Severson, R. K., Stevens, R. G., Kaune, W. T., Thomas, D. B., Heuser, L., Davis, S., and Sever, L. E., Acute non-lymphocytic and residential exposure to power frequency magnetic fields, *Am. J. Epidemiol.*, 128, 10, 1988.

Shaw, G. M. and Croen, L. A., Human adverse reproductive outcomes and electromagnetic field exposures: review of epidemiologic studies, *Environ. Health Perspect.*, 1993, 101 (Suppl. 4), 107, 1993.

Speers, M. A., Dobbins, J. G., and Miller, V. S., Occupational exposures and brain cancer mortality: a preliminary study of East Texas residents. *Am. J. Industr. Med.*, 13, 629, 1988.

Stern, F. B., Waxweiler, R. J., Beaumont, J. J., Lee, S. T., Rinsky, R. A., Zumwalde, R. D., Halperin, W. E., Bierbaum, P. J., Landrigan, P. J., and Murray, Jr., W. E., A case-control study of leukemia at a naval nuclear shipyard, *Am. J. Epidemiol.*, 123, 980, 1986.

Stevens, R. G., Davis, S., Thomas, D. B., Anderson, L. E., and Wilson, B. W., Electric power, pineal function, and the risk of breast cancer, *FASEB J*, 6, 853, 1992.

Stollery, B. T., Effects of 50 Hz electric currents on mood and verbal reasoning skills, *Br. J. Ind. Med.*, 43, 339, 1986.

Theriault, G. P., Health effects of electromagnetic radiation on workers: epidemiologic studies, in *Proceedings of the Scientific Workshop on the Health Effects of Electric and Magnetic Fields on Workers.* Bierbaum, P. S., Savitz, D, A., and Peters, J. M., Eds., U.S. DHHS, National Institute for Occupational Safety and Health, Cincinnati, Ohio, 1991, p. 23.

Tomenius, L., 50-Hz electromagnetic environment and the incidence of childhood tumors in Stockholm county, *Bioelectromagnetics*, 7, 191, 1986.

Vena, J. E., Graham, S., Hellmann, R., Swanson, M., and Brasure, J., Use of electric blankets and risk of postmenopausal breast cancer, *Am. J. Epidemiol.*, 134, 180, 1991.

Verreault, R., Weiss, N. S., Hollenbach, K. A., Strader, C. H., and Daling, J. R., Use of electric blankets and risk of testicular cancer, *Am. J. Epidemiol.*, 131,759, 1990.

Wertheimer, N. and Leeper, E., Electrical wiring configurations and childhood cancer, *Am. J. Epidemiol.*, 109, 273, 1979.

Wertheimer, N. and Leeper, E., Re: Electrical wiring configurations and childhood leukemia in Rhode Island (Letter), *Am. J. Epidemiol.*, 111, 461, 1980.

Wertheimer, N. and Leeper, E., Adult cancer related to electrical wires near the home. *Int. J. Epidemiol.*, 11, 345, 1982.

Wertheimer, N. and Leeper, E., Fetal loss associated with two seasonal sources of electromagnetic field exposure, *Am. J. Epidemiol.* 129, 220, 1989.

Wong, O. and Decoufle, P., Methodological issues involving the standardized mortality ratio and proportionate mortality ratio in occupational studies, *J. Occup. Med.*, 24, 299, 1982.

Youngson, J. H. A. M., Clayden, A. D., Myers, A., and Cartwright, R. A., A case/control study of adult haematological malignancies in relation to overhead powerlines, *Br. J. Cancer*, 63, 977, 1991.

8 MAN-MADE CHEMICAL DISASTERS

Pier Alberto Bertazzi

CONTENTS

INTRODUCTION

In 1990, the 44th General Assembly of the United Nations launched the decade for the reduction of frequency and impact of natural disasters.[15] A committee of experts endorsed a definition of disasters as "a disruption of the human ecology that exceeds the capacity of the community to function normally". This definition is general enough to include both natural and man-made disasters. Among the latter, of special relevance and concern nowadays,

are major industrial accidents defined by a European Communities Council
Directive as "an occurrence such as a major emission, fire or explosion resulting from uncontrolled developments in the course of an industrial activity,
leading to a serious danger to man, immediate or delayed, inside or outside
the establishment, and/or to the environment, and involving one or more
dangerous substances".[12]

Numerous and different definitions and classifications of disasters are
available and have been recently reviewed.[18,30,35,54] Two of them are mentioned
here as examples. The U.S. Centers for Disease Control[9] identified three major
categories of disasters: geographical events such as earthquakes and volcanic
eruptions; weather-related problems including hurricanes, tornadoes, heat
waves, cold environments, and floods; and, finally, human-generated problems
which encompass famine, air pollution, industrial disasters, fires, and nuclear
reactor incidents. Another classification by cause[43] included weather and geological events among natural disasters, whereas man-made causes were characterised as non-natural, technological, purposeful events perpetuated by man
(e.g., transportation, war, fire/explosion, chemical and radioactive release).

The remainder of this discussion will only focus on industrial disasters
or major accidents as defined by the European Economic Community (EEC).
Radiation and nuclear reactor accidents are the subject of other chapters.

INDUSTRIAL CHEMICAL DISASTERS

Disasters of industrial origin share common features, yet are different in
their mode of occurrence and consequences.[6] Most industrial disasters occur
through a sudden, dramatic incident which leaves no ambiguity about the
serious hazard being posed to the surrounding environment and population.
Typical examples of "overt" disasters of this type are

1. The runaway reaction in the vessel for the production of trichlorophenol
   in the ICMESA plant, near Seveso, Italy that caused a release of chemicals,
   including 2,3,7,8-TCDD (dioxin), over several square kilometres of popu-
   lated countryside[23,44]
2. The runaway reaction that caused the pressure-release valve of the methyl
   isocyanate (MIC) storage tank of a pesticides plant to burst open, spreading
   a deadly cloud over the city of Bhopal, central India[13,40]
3. The fire that burned down a storehouse of chemical substances (mainly
   agricultural chemicals) in Schweizerhalle, near Basel, Switzerland, causing
   the spill of a huge amount of chemicals along with water from the fire
   extinguishers into the Rhine[1]

Other industry-linked disasters may occur, however, in a diluted, silent
manner and may become apparent only when, as time passes, the human and/or

environmental targets on the path of the "covert" release or spill show unequiv-
ocal signs of damage. Thus, for example, it took several years to identify alkyl
mercury as the etiologic agent of the "Minamata disease" and to uncover the
source of the toxic release in a factory discharging its effluents into the sea.[24]
Similar harm to human health and environmental salubrity may originate from
waste disposal sites. Love Canal, near Niagara Falls, U.S., represents the
prototypical example[20] of the hazard possibly entailed by the improper dumping of toxic chemicals.[7,19] Another example of this type of disaster comes from
the many episodes of mass poisonings due to food contaminated by toxic
chemicals. Reviews on these episodes are available.[18,54] The most recent mass
poisonings of this type occurred in Japan,[37] Taiwan,[10] and Spain.[55]

One further feature of the mode of occurrence of today's industrial disasters is transnationality. Apart from Chernobyl, which caused radiological contamination on a continental scale, one can easily note, for example, that at
Schweizerhalle the chemical spill into the Rhine potentially affected four
European countries (Switzerland, France, Germany, and The Netherlands).
Transnationality applies not only to consequences of industrial disasters, but
also to (remote) causes. Thus, for instance, some people noted that the Bhopal
disaster probably occurred because of specific acts and decisions taken some-
where in the corporate superstructure of the multinational, but not in Bhopal.[16]

This brings us to another characteristic of industrial disasters that can be
expected in the future: their prevailing occurrence in developing countries.[26,28]
Hazardous industrial activities, as well as waste materials, are increasingly
exported to world regions where the least stringent protection measures exist.
In addition, the emerging pattern of industrialisation and the modernisation of
agriculture in those same countries involve the application and use of imported
or adopted technology within contexts quite different from those for which
they were intended. Furthermore, industrial activities frequently become concentrated in overcrowded urban settlements with a shortage of community
services. Such a setting could clearly aggravate the consequences of any
industrial disaster.

One final common characteristic of these major toxic releases is that they
are a preventable cause of disaster. Although this is not just a matter of health
sciences, epidemiology may serve this task in different ways.

DISASTERS AS DISEASE DETERMINANTS

The briefly outlined characteristics of industrial disasters vividly convey
the vastness, variety, and complexity of the factor "disaster" as a determinant
of disease occurrence in the population. At least three main components can
be singled out.

The Release

Nature and Entity

The nature of the event is by no means obvious, not just in the case of subtle, unobserved environmental contamination, but even when the disaster is overt. Thus, in the case of Seveso, it took ten days to clarify which toxic compound was concerned;[49] moreover, 15 years after the accident, the quantity of 2,3,7,8-tetrachlorodibenzo-para-dioxin (TCDD) released is still a matter of dispute.[14] By the same token, medical observations, as well as autopsy findings on Bhopal victims raised doubts about the presence of other toxic chemicals in addition to MIC in the release.[48,52]

Context

The ecological, social, political, and cultural context in which the disaster occurs might well override the magnitude of the release in relevance. In Bhopal, for example, over the years the plant was in operation, the city had expanded up to the boundaries of the plant; and several hundred thousand people were living in crumbling houses providing no shelter against the contamination, in narrow and poorly illuminated streets, with no transportation or adequate communication, etc.[27,43] In addition, the time of year and climate might also play a major role in determining exposure.

Individual Disaster Experience

Irrespective of the toxicity of the compounds involved, the event of an explosion, fire, or chemical spill, etc., always carries a burden of fear, anxiety, and distress for the population involved. Increasing attention is being paid to psychosocial aspects as determinants of post-disaster health effects.[21,39,45,46,50,51] Serious social and psychological consequences have been documented in the Love Canal[22] and Seveso[6] populations. Risk factors pertaining to the disaster-associated stress stem from either objective losses/experiences or the perception/evaluation of the disaster.[35] Industrial disasters often produce a high level of perceived risk, even though objective losses are few. Several surveys have ranked chemicals and nuclear power highest among perceived hazards.[31]

Response measures

Rhine pollution at Schweizerhalle was caused by a response measure (i.e., fire extinguishing). The example is extreme, but it is useful to illustrate how response measures might be among the most relevant determinants of health effects of disasters. They can alleviate or aggravate the consequences for the exposed population. For example, there is no doubt that evacuating people from their contaminated home and land prevents further exposure; yet, for many people, this may represent an additional and possibly very severe risk factor for further adverse health effects.[32] Another component of the response which has been shown to have modified the effects of disasters on individuals is social support.[35]

Health Outcomes

All the above factors ought to be taken into consideration as possible predictors of the adverse effects of disaster occurrences. Accordingly, health outcomes associated with disasters are numerous and diverse in nature and time of appearance.

Toxic Effects

These are principally related to the toxicokinetics and toxicodynamics of the released chemicals, and to the factors (host susceptibility, concomitant exposures, etc.) capable of modifying them. Toxic effects can be categorised as:

- Immediate (e.g., caustic dermal lesions in Seveso children[8])
- Early (e.g., chloracne in Seveso people[2])
- Long-term (e.g., lymphoma in firefighters years after acute exposure[36])

The occurrence of specific effects may actually help in identifying the nature of the toxic agent released. This was the case with the type of neurologic syndrome observed in Minamata.[24]

Stress-Related Effects

These effects are linked to the disastrous/catastrophic nature of the event as experienced by the population. They can be both mental and physical.

Mental consequences are not necessarily immediate (e.g., anxiety, fear, etc.); delayed (months or years) responses have been documented, but are often overlooked.[21,34,38,45,51] The term "post-traumatic stress disorder" was introduced in 1980 into psychiatric nomenclature following repeated observations of mental sequelae in survivors of various types of disaster/catastrophe.[33,53]

Physical consequences have been, for example, uncovered in Seveso, with the unanticipated finding of an increased mortality from cardiovascular diseases in the early post-accident period; this occurrence most probably is due to the precipitation, under severe stress experience, of acute episodes in persons with pre-existing ill health.[5]

RESEARCH NEEDS

From the epidemiologic standpoint, research needs are related to the following research goals:

- Identification of disaster sources and disaster-mediating factors
- Estimation of health impact (early and delayed)
- Evaluation of effectiveness of post-disaster measures

These epidemiologic achievements can contribute to the more general task of disaster prevention, emergency preparedness, and emergency response and relief, which also require the active participation of experts and institutions from outside the scientific community.

Identification of disaster sources

Information on potential sources of chemical disasters should be collected, updated, and made available to the scientific community at the national and international level. The EEC definition of major accidents might be useful to specify the object and delimit the field. The long and rich experience in occupational settings may serve the purpose of identifying point sources of potential disastrous impact for the population and the environment. Existing data bases on chemical accidents (e.g., MHIDAS[11]) should be made more widely known and accessible, possibly through existing international bodies.

Once a potential source of chemical disaster is identified, the health experience of the population living in the area concerned should become the object of a continuous surveillance by means of existing health records (vital statistics, death certificate, cancer registration, records on other health related events). Small area statistics methods are particularly relevant to the achievement of this goal. This surveillance system might be particularly useful to uncover "silent" or "diluted" disastrous impact of the source on the population, but also represents an element of preparedness in case of disastrous incident.

The availability of this information and a surveillance system is of paramount importance in developing countries, where, on the contrary, one can easily anticipate that the worst difficulties will be encountered.[25] Transnational siting policy for hazardous establishments should take this need into consideration. Emergency plans for communities around industrial installations have been prepared in several industrialised countries as, for example, the U.K.[3] They also include requirements for population identification; long-term epidemiologic studies; and community information, awareness, and preparedness. When we export and adapt industrial installations, we should in parallel export and adapt prevention and emergency plans and resources.[41]

Disaster-Mediating Factors

The type of chemical hazard is by no means the only predictor of the potential disastrous impact. All other intervening and mediating factors should be identified and evaluated. They include location of the source with respect to residential sites and other environmental targets possibly relevant to human exposure (food chain, water supply, etc.); structure of the residential community, including type of housing and availability of services; community information and preparedness; emergency plans and response strategies; actions taken by residents; and compliance of individuals in the community with official orders. A comprehensive literature review may be a first step to uncover and, possibly, single out the role of at least some of these mediating factors. In case sufficient and valid data can be retrieved, case-control studies within the disaster community (cohort) should be conducted to unravel which hypothesised factors were actually relevant in mediating the occurrence of adverse health effects.

Estimation of Health Impact

Three main research requirements deserve priority consideration:

1. Individual exposure characterisation and quantification
2. Methods and tools for the long-term surveillance of individuals
3. Health outcomes ascertainment

Exposure ascertainment may be based on measurements of the types and concentrations of chemicals in various environmental media (air, soil, water, vegetation, farm animals, wildlife, food, etc.). A questionnaire can help in reconstructing the opportunities for exposure for every individual subject in the post-disaster period. Biological samples are the only faithful means to document individual exposure. The storage of biological samples should be planned whenever possible, and might particularly prove rewarding. Advanced analytical capabilities or new types of assays might, in fact, become available at a later stage; and might allow a more accurate evaluation of exposure, a prediction of delayed effects, and an ascertainment of individual susceptibility.

In the hectic social climate of an emergency phase, the validity of exposure assessment may be hampered by several circumstances and factors, such as, for example, scarcity of resources, time constraint, political pressure, etc. Usually, different labs and teams are called in during the emergency phase and, in a disaster scenario, issues of validity and quality control might receive little attention.

A great deal of information is available on exposure to toxic chemicals in occupational settings, whereas less is known on their environmental fate, routes of exposure outside workplaces, persistence and accumulation in eco-systems, and effects on biota. This knowledge is essential for the ascertainment of potential human exposure. Environmental toxicology studies are increasingly needed to cope with current problems of environmental pollution, and their dramatic magnification in case of disasters. The suitability and feasibility of biological monitoring[29] in the general population cannot be automatically inferred by the experience gained in working populations, especially when one considers that in the general environment susceptible persons, such as children, elderly, and ill people might be exposed, and for longer than a workshift.

The relevant population is composed of all those potentially exposed, irrespective of the presence of immediate damage. Generally, the identification of the exposed population rests on the spatial extent of the contamination. Permanent residents are in this way included, provided that a valid and comprehensive roster exists (this was not the case in Bhopal). In specific situations, additional means other than the roster of residents may allow a more accurate estimate of the size of exposed population.[17] In any disaster setting there are groups of nonresident people who can be identified as probably exposed. They include, for example, people involved in emergency or postemergency activities, postdisaster field workers, and reclamation workers. They should be enumerated and included in postdisaster surveillance programs. People who present immediate, specific signs or symptoms are readily identifiable as disaster-exposed, even if not resident. People move, and especially after a disastrous occurrence those who move might well be the most relevant to the ascertainment of health effects (especially the delayed ones). Therefore, it is essential that good vital statistics information exists so that an active mechanism of population identification, enumeration, and follow-up can be implemented. A complete census of the population in the area is of paramount importance even for the purpose of identifying suitable control groups. In fact, often only controls from the area allow proper consideration of variables acting as powerful confounders or modifiers.[42]

It is apparent, from previous disaster experience, how difficult it is to find a remedy for the inadequacy of vital statistics data in the postimpact phase and for the lack of data base and information systems suitable for the enumeration and follow-up of the population in the affected area. If such resources do not exist on a routine basis, they should be established as part of the siting policy for potentially dangerous establishments. After the chemical release, toxic and mental effects, and short- and long-term outcomes should be anticipated. A long-term follow-up should be conducted on at least a sample of subjects involved in disasters, since delayed effects of environmental exposure to chemicals have been investigated in very few instances, and the psychosocial impact of being involved in a chemical disaster has still to be satisfactorily evaluated. The sample can consist of the group of cases of a specified illness in a given period of time, and a sample of comparable subjects free of illness in the same period.

In general, the long-term follow-up of a population helps to discriminate between the "natural" history of chemical-related effects and the "natural" history of psychosomatic effects related to the stress experience caused by the incident itself or by other intervening variables, (e.g., political and communication issues), and to evaluate whether and how they interact.

Evaluation of Post-Disaster Measures

This evaluation is largely made possible by the achievement of the above research goals. A first evaluation concerns methods, procedures and tools for successful early case ascertainment. This phase is crucial since the allocation of services, treatment and rehabilitation, and planning of future action depend on it.[4]

Long-term surveillance is the most rewarding means for the evaluation of post-emergency measures. The surveillance should be concerned with both persisting exposure opportunities, and late appearance of unexpected or unanticipated events. Two research approaches can be pursued in populations already exposed to disaster. One is the comparison of disease occurrence in subgroups of the population known to have or not have been exposed to specified measures. The other one is based on a case-control approach and requires comparison of exposure to post-disaster measures (e.g., evacuation, medical/preventive treatment, social support, etc.) among persons with diseases potentially linked to the chemical release, and comparable people without the disease. A "sentinel events" approach[47] may also be useful to reveal the persistence of a threat when anticipated effects of exposure are quite specific.

In summary, when a place is known as a putative major accident source, it should become first a privileged site of technical and social primary prevention actions, and second a privileged source of valid and comprehensive information capable of guiding action in case of disaster occurrence.

The information recorded in previous disastrous accidents should be reconsidered in a synoptic way to evaluate the relevance and role of the aforementioned variables, and the success of actions taken. Should existing data allow it, meta-analysis of the homogeneous type of disaster occurrences would prove most useful for evaluation purposes. This could generate guidelines for an epidemiologic approach to chemical disasters.

Finally, for the success of the contribution epidemiology can give to the ascertainment, relief, and prevention of chemical disaster effects, some organisational and feasibility requisites are particularly relevant and are listed here as concluding suggestions.

- Different experts should be on the scene in the immediate aftermath of a disaster, working in close coordination. They should cover the fields related to the environmental fate of the agent, its toxic properties to humans and biota, analytical methods, clinical medicine and pathology, biostatistics, and epidemiology.
- Based on preexisting and/or early available evidence, a comprehensive study plan should be developed as early as possible to identify goals, problems, and resource requirements.
- Early phase activities affect the course of any subsequent action. Since long-term effects should be expected after virtually every type of industrial disaster, great care should be devoted to ensure availability of information for potential future study.
- Feasibility should be given high consideration to facilitate scientific and public health achievements and clarity of communication.
- For reasons of validity and cost-effectiveness, it is advisable to rely on "hard" information whenever available, either in identifying and enumerating the study population or in estimating exposure and choosing the study endpoints.
- International cooperation ought to be strengthened in the domains of health science, public health, and industrial siting and safety policy.

It would be advisable to have a group of experts "on-call" under WHO coordination to provide rapid assistance to countries in need.

REFERENCES

1. Ackermann-Liebrich, U. A., Braun, C., and Rapp, R. C., Epidemiologic analysis of an environmental disaster: the Schweizerhalle experience. *Environ. Res.*, 58, 1, 1992.
2. Assennato, G., Cervino, D., Emmett, E. A., Longo, G., and Merlo, F., Follow-up of subjects who developed chloracne following TCDD exposure at Seveso. *Am. J. Ind. Med.*, 16, 119, 1989.
3. Baxter, P. J., Davies, P. C., and Murray V., Medical planning for toxic releases into the community: the example of chlorine gas. *Br. J. Ind. Med.*, 46, 277, 1989.
4. Baxter, P. J., Responding to major toxic releases. *Ann. Occup. Hyg.*, 34, 615, 1990.
5. Bertazzi, P. A., Zocchetti, C., Pesatori, A. C., Guercilena, S., Sanarico, M., and Radice, L., Ten-year mortality study of the population involved in the Seveso incident in 1976. *Am. J. Epidemiol.*, 129, 1187, 1989.
6. Bertazzi, P. A., Industrial disasters and epidemiology. A review of recent experiences. *Scand. J. Work Environ. Health*, 15, 85, 1989.
7. Buffler, P. A., Crane, M., and Key, M. M., Possibilities of detecting health effects by studies of populations exposed to chemicals from waste disposal sites. *Environ. Health Perspect.*, 62, 423, 1985.

8. Caramaschi, R., Del Corno, G., Favaretti, C., Giambelluca, S. E., Montesarchio, E., and Fara, G. M., Chloracne following environmental contamination by TCDD in Seveso, Italy. *Int. J. Epidemiol.*, 10, 135, 1981.

9. Centers for Disease Control. *The Public Health Consequences of Disasters 1989*. Centers for Disease Control, Atlanta, GA, 1989.

10. Chen, P. S., Luo, M. L., Wong, C. K., and Chen, C. J., Polychlorinated biphenyls, dibenzofurans, and quaterphenyls in toxic rice-bran oil and PCBs in the blood of patients with PCB poisoning in Taiwan. *Am. J. Ind. Med.*, 5, 133, 1984.

11. Clifton, J. J. and Wilkinson, A., Major hazard incident data service. In: Istituto Superiore Sanità, World Health Organization, International Programme on Chemical Safety. World Conference Chemical Accidents. CEP Consultants Ltd., Edinburgh 1987, 64.

12. Council of the European Commission. Council Directive of 24 June 1982 on the major-accident hazards of certain industrial activities. Official Journal of the European Commission, No. L 230, 5/8/1982, 1.

13. Das, J. J., The Bhopal tragedy. *J. Indian Med. Assoc.*, 83, 72, 1985.

14. Di Domenico, A., Cerlesi, S., and Ratti, S., A two-exponential model to describe the vanishing trend of 2,3,7,8-tetrachlorodibenzo-p-dioxin (TCDD) in the soil at Seveso, Northern Italy. *Chemosphere*, 20, 1559, 1990.

15. Editorial. Disaster epidemiology. *Lancet*, 336, 845, 1990.

16. Friedrich Naumann Foundation. Industrial hazards in transnational world: risk, equity and empowerment. Council on International and Public Affair, New York, NY 1987, 13.

17. Glickman, T. S., A methodology for estimating time-of-day variations in the size of a population exposed to tisk. *Risk Anal.*, 6, 317, 1986.

18. Grisham, J. W., Ed. *Health Aspects of the Disposal of Waste Chemicals*. Pergamon Press, New York, 1986, 220.

19. Heath, C. W. Jr., Field epidemiologic studies of populations exposed to waste dumps. *Environ. Health Perspect.*, 48, 3, 1983.

20. Heath, C. W. Jr., Nadel, M. R., Zack, M. M. Jr., Chen, A. T. L., Bender, M. A., and Preston J., Cytogenetic findings in persons living near the Love Canal. *JAMA*, 251, 1437, 1984.

21. Hodgkinson, P. E., Technological disaster. Survival and bereavement. *Soc. Sci. Med.*, 29, 351, 1989.

22. Holden, C., Love Canal residents under stress. *Science*, 208, 124, 1980.

23. Homberger, E., Reggiani, G., Sambeth, J., and Wipf, H. K., The Seveso accident: its nature, extent and consequences. *Ann. Occup. Hyg.*, 22, 327, 1979.

24. Hunter, D., *The Diseases of Occupations*. Hodder and Stoughton, London, 1978, 337.

25. Jasanoff, S., The Bhopal disaster and the right to know. *Soc. Sci. Med.*, 27, 1113, 1988.

26. Jeyaratnam, J., 1984 and occupational health in developing countries. *Scand. J. Work Environ. Health*, 11, 229, 1985.

27. Koplan, J. P., Falk H., and Green G., Public health lesson from the Bhopal chemical disaster. *JAMA*, 264, 2795, 1990.

28. LaDou, J., Deadly migration. Hazardous industries' flight to the third world. *Technol. Rev.*, July 1991, 47.

29. Lauwerys, R. R., *Industrial Chemical Exposure: Guidelines for Biological Monitoring*. Biomedical Publications, Davis, CA, 1983.

30. Lechat, M. F., The epidemiology of health effects of disasters. *Epidemiol. Rev.*, 12, 192 1990.

31. Lee, T. R., Public attitudes towards chemical hazards. *Sci. Total Environ.*, 51, 125, 1986.

32. Levenson, M. and Rahn, F. J., Is evacuation the best policy? *Science*, 208, 131, 1980.

33. Lindy, J. D., Green, B. L., and Grace, M. C., The stressor criterion and posttraumatic stress disorder. *J. Nerv. Ment. Dis.*, 175, 269, 1987.

34. Logue, J. N., Melick, M. E., and Struening, E., A study of health and mental health status following a major natural disaster. In: *Research in Community and Mental Health: an Annual Compilation of Research. Volume 2*. Simmons, R., Ed., JAI Press, London, 1981, 217.

35. Logue, J. N., Melick, M. E., and Hansen, H., Research issues and directions in the epidemiology of health effects of disasters. *Epidemiol. Rev.*, 3, 140, 1981.

36. Markovitz, A. and Crosby, W. H., A soil fumigant, 1,3-dichloropropene, as possible cause of hematologic malignancies. *Arch. Intern. Med.*, 144, 1409, 1984.

37. Masuda, Y. and Yoshimura, H., Polychlorinated biphenyls and dibenzofurans in patients with Yusho and their toxicological significance: a review. *Am. J. Ind. Med.*, 5, 31, 1984.

38. McFarlane, A. C., Posttraumatic morbidity of a disaster. A study of cases presenting for psychiatric treatment. *J. Nerv. Ment. Dis.*, 17, 4, 1986.

39. Melick, M. E., Logue, J. N., and Frederick, C. J., Stress and disaster. In: *Handbook of Stress. Theoretical and Clinical Aspects*. Goldberger, L., and Breznitz, S., Eds., Free Press, New York, 1982, 613.

40. Metha, P. S., Metha, A. S., Metha, S. J., and Makhijani, A. B., Bhopal tragedy's health effects: a review of methyl isocyanate toxicity. *JAMA*, 264, 2781, 1990.

41. Misra, V., Jaffrey, F. N., and Viswanathan, P. N., Risk analysis in hazardous industries. *Regul. Toxicol. Pharmacol.*, 13, 62, 1991.

42. Neutra, R., Roles for epidemiology: the impact of environmental chemicals. *Environ. Health Perspect.*, 48, 99, 1983.

43. Parrish, R. G., Falk, H., and Melius, J. M., Industrial disasters: classification, investigation, and prevention. In: *Recent Advances in Occupational Health*. Harrington, J. M., Ed., Churchill Livingstone, Edinburgh, 1987, 155.

44. Pocchiari, F., Di Domenico, A., Silano, V., and Zapponi, G., Environmental impact of the accidental release of tetrachlorodibenzo-p-dioxin (TCDD) at Seveso. In: *Accidental Exposure to Dioxins. Human Health Aspects*. Coulston, F., and Pocchiari, F., Eds., Academic Press, New York, 1983, 5.

45. Raphael, B., Mental and physical health consequences of disasters. *Med. J. Aust.*, 143, 180, 1985.

46. Reko, K., The psychosocial impact of environmental disasters. *Bull. Environ. Contam. Toxicol.*, 33, 655, 1984.

47. Rutstein, D. D., Mullan, R., Frazier, T. M., Halperin, W. E., Melius, J. M., and Sestito, J. P., Sentinel health events (occupational): a basis for physician recognition and public health surveillance. *Am. J. Public Health*, 73, 1054, 1983.

48. Salmon, A. G., Bright red blood of Bhopal victims: cyanide or MIC? *Br. J. Ind. Med.*, 43, 502, 1986.

49. Silano, V., Case study: accidental release of 2,3,7,8-tetrachlorodibenzo-p-dioxin (TCDD) at Seveso, Italy. In: Planning Emergency Response Systems for Chemical Accidents. Copenhagen WHO Regional Office for Europe, 1981, 167.

50. Singer, T. J., An introduction to disaster: some considerations of a psychological nature. *Aviat. Space Environ. Med.*, 53, 245, 1982.

51. Singh, B. S., The long-term psychological consequences of disaster. *Med. J. Aust.*, 145, 555, 1986.

52. Sriramachari, S., Pathology collection and storage of pathologic material for immediate analysis and later study of toxicological effects and their long-term implication. Presented at 6th Workshop of the Scientific Group on Methodologies for the Safety Evaluation of Chemical (SGOMSEC), New Delhi, January 27-February 2, 1987.

53. Ursano, R. J., Posttraumatic stress disorder: the stressor criterion. *J. Nerv. Ment. Dis.*, 175, 273, 1987.

54. Weiss, B. and Clarkson, T. W., Toxic chemical disaster and the implication of Bhopal for technology transfer. *Milbank Mem. Fund Q.*, 64, 216, 1986.

55. World Health Organization. Toxic oil syndrome. Mass food poisoning in Spain. Copenhagen WHO Regional Office for Europe, 1984, 3-15.

PART 2: DISEASE

9 NEOPLASTIC DISEASES (LUNG CANCER EXCLUDED)

Lorenzo Simonato

CONTENTS

INTRODUCTION

Cancer has been a disease whose increasing frequency has negatively characterised the history of human health during this century. This phenomenon was paralleled by the success of modern societies in fighting against diseases which have been major causes of death such as tuberculosis and other infectious diseases. Cancer prevention and control remains the main target for public health; and it is therefore essential for research, and epidemiological research, in particular, to harmonize its priorities with the other strategies.

This chapter will briefly review the available epidemiological evidence for sites other than lung, eventually proposing some priorities for epidemiological research. After excluding lung cancer there are about 60 other malignant neoplasms and another group of about 30 entities, including benign and undefined tumours. This chapter will necessarily be limited to some selected sites and will not obviously be exhaustive of the available knowledge on all cancer sites.

EPIDEMIOLOGICAL EVIDENCE ON SELECTED CANCER SITES

Oral Cavity (ICD IX 140–149)

Cancer of the oral cavity is characterized by a large geographical variability and has causally been associated with personal habits, such as alcohol drinking and tobacco smoking and chewing.

Time trends indicate a moderate increase in most Western countries (IARC, 1990; Higginson et al., 1992), while the frequency of this group of tumours is decreasing in some of the countries where the rate was very high (Jayant and Yeole, 1987).

Oesophagus (ICD IX 150)

Oesophageal cancer is also characterized by a large geographical variability and by a slight tendency to increase over time. Cancer of the oesophagus has been causally associated with alcohol drinking and tobacco consumption

(Tuyns, 1982) and has been shown to be more frequent in populations with poor nutritional intake (Day and Muñoz, 1982; Victora et al., 1987) and possibly in different races (Blot and Fraumeni, 1987; Bang et al., 1988).

Stomach (ICD IX 151)

Cancer of the stomach is one of the most common form of cancer in the world and is characterized by a large geographical variability and by a steady decline during the last five decades in most countries (Muñoz, 1988).

The etiopathogenesis of this tumour has been associated with exogenous factors mainly of dietary origin (La Vecchia et al., 1987; Hirayama, 1988; Buiatti et al., 1989; Graham et al., 1990). The possible role of air pollution can be found in WHO reports and other papers. There is consistent, but not definitive, evidence of an increased risk from high intake of salty and smoked food, fried food, complex carbohydrates, and starchy food; and of a protective role of a diet rich in fruits and vegetables.

Large Intestine (ICD IX 153–154)

Colorectal cancer is more frequent in developed countries, where it is relatively stable, while it is increasing in countries or in ethnic groups where the rates were usually low.

The epidemiological evidence indicates exogenous factors, probably of dietary origin, as the most probable causes. The role of drinking water is discussed elsewhere (refer to Chapter 5 in this volume)

Several studies (Cannon-Albright et al., 1988; Bodmer et al., 1987; Fearon et al., 1990) have suggested the possible role of hereditary factors and individual susceptibility.

Liver (ICD IX 155)

Cancer of the liver is increasing in many countries, although it is particularly difficult to make comparisons over time for this tumour, due to the changes in the International Classification and of diagnostic criteria (Muñoz and Bosch, 1987).

Epidemiological evidence has consistently associated high risk of liver cancer with hepatitis B virus (HBV) infection, aflatoxin food contamination, and elevated consumption of alcohol. Vinyl chloride monomer has been established to be the cause of liver angiosarcoma (LAS) in occupational populations.

Pancreas (ICD IX 157)

Pancreatic cancer is reported as an increasing disease in most countries and it is more common in developed countries. It is not one of the most frequent cancers, but it is rapidly fatal. Tobacco smoking is the most important risk factor so far identified (Cuzick and Babiker, 1989), while other dietary factors like alcohol and coffee drinking do not appear consistent (Boyle et al., 1989).

Larynx (ICD IX 161)

Laryngeal cancer is predominantly a male's disease. It is relatively frequent in South America and in Mediterranean countries. Time trends are not consistent with the risk of lung cancer. Tobacco smoking and alcohol drinking are the two major risk factors for this tumour acting in a interactive way when both present (Tuyns et al., 1988). Different mechanisms of action of tobacco smoking on different sites is suggested.

Nickel refining is the most consistent occupational factor which has been causally associated with larynx cancer (Brown et al., 1988).

Soft Tissue Sarcoma (ICD IX 171)

The frequency of this tumour is around 2–3 cases per 100,000 person-years. Although not being very frequent, time trends indicating an apparent increase in many countries draw attention to this group of neoplasms.

Several chemical exposures have been associated with soft tissue tumours mostly of occupational origin, such as phenoxyacetic acids and chlorophenols.

The link between Kaposi's sarcoma and AIDS is now established. Medical use of Thorotrast has also been shown to increase the risk of developing LAS.

Skin Melanoma (ICD IX 172)

This relatively rare tumour is reported as an increasing trend in most countries with two different features. While it is increasing slowly in deeply pigmented populations which are affected by high rates, annual increases in the order of 5–7% per annum have been reported in North America and in Nordic countries in Europe.

Sun-induced nevi seem to be a predisposing factor to the risk of developing malignant melanoma (Osterlind et al., 1988a). Intense exposure to sunlight resulting in sunburns appears to be the most important cause of malignant melanoma (Osterlind et al., 1988b; Elwood et al., 1987; IARC, 1992) indicating ultra violet light as the most important component. Increased risk was also found for users of sunlamps at home (Walter, 1990), while the possible effects from fluorescent lamps are still under debate (Béral et al., 1982).

Breast (ICD IX 174)

Cancer of the breast is the most common tumour among women with more than 500,000 cases estimated in 1980 (Parkin, 1988). It represents about one fourth of all cancer cases in high risk areas, like North America and

Western Europe; while, in low risk areas like China and Japan, it accounts for 1 in 4 to 1 in 16 of all cancer cases.

Incidence rates have been increasing in most countries during the last four or five decades and the risk has been more marked in post-menopausal women (Hakulinen et al., 1986) with the exception of Singapore. Mortality during the same period has been rather stable or characterised by a moderate increase.

The major risk factors are related to reproductive habits and diet. Tobacco smoking has also been associated with an increased risk of breast cancer, but the evidence is not consistent. High levels of alcohol consumption can also be associated with increased risk of breast cancer.

Uterine Cervix (ICD IX 180)

Cervical cancer is the second most frequent cancer among women accounting for 465,600 estimated new cases in 1980 (Parkin, 1988), the majority of which (80%) occurring in developing countries.

Incidence rates are generally higher in urban than in rural populations. Time trends indicate a generalized decline of both incidence and mortality in most countries, where screening programmes have been introduced with the exception of mortality among young women in U.K., Australia, New Zealand, and Norway (Muñoz and Bosch, 1989). High risk areas are South America, sub-Saharan Africa, and South East Asia, while a low risk area is the Middle East zone.

Risk factors for cervical cancer have been related to sexually transmitted infections, high parity, and inadequate medical services. Human papilloma virus infection has also been associated with increased risk, and the epidemiological data appear consistent with a causal interpretation (IARC, 1992).

Corpus Uteri (ICD IX 182)

The frequency of cancer of the uterine corpus is about one third of that of the cervix with 149,000 new estimated cases in 1980 (Parkin, 1988).

Incidence rates are variable and an apparently increasing trend in post-menopausal women is present in U.S. and in some Nordic countries, while mortality rates have been steeply declining in most countries mainly due to the improved survival.

Endometrial cancer is associated with obesity, and relating to factors suggesting the important role of endogenous hormones.

Ovary (ICD IX 183)

Ovarian cancer represents the highest mortality from gynaecological tumours. Rates are higher in Northern and Western Europe and in North America, and low in Japan, China and India.

Rates seem to be rather stable over time or characterized by moderate increases. Little is known about the etiology of this tumour, which is probably influenced by endogenous and exogenous steroid hormones associated with reproductive behaviour.

Prostate (ICD IX 185)

Prostatic cancer is an important disease due to longer life expectancy, and constitutes the most common tumour among elderly males. The rates are higher among the members of high social class in U.K. as compared to lower class.

In most population incidence rates are rapidly increasing, partly due to the improvement of diagnostic criteria and to the consequent more frequent report of latent cancers. Mortality is also increasing but not as rapidly as incidence in many countries, with the exception of U.S., U.K., and Australia, where mortality is rather stable.

Diet has been associated with higher risk of prostatic cancer (Zaridze and Boyle, 1987), as has oestrogen and testosterone stimulation as well as sexual activity (Hill et al., 1982; Ross et al., 1987).

Occupational exposure to cadmium and smoking has frequently been associated with an increased risk of prostate cancer.

Bladder (ICD IX 188)

Cancer of the urinary bladder is more common among males and exhibits higher rates in whites in North America and North Western Europe. Racial and geographical differences are important.

Time trends show a generalized increase both in incidence and mortality, although the pattern is more moderate for the latter.

Major determinants of bladder cancer are some occupational exposures, tobacco, and parasitic infestations.

Since the 1950s, epidemiological studies have demonstrated the carcinogenic effects of a number of aromatic amines on exposed workers in dyes manufacturing and rubber industry. Other excesses of bladder cancer have been reported among workers employed in the leather industry, painters, truck drivers, and other occupational groups for which the evidence, however, is not definitive.

Tobacco smoking is a proven cause of bladder cancer and the risk is higher for users of black tobacco, as compared to blond tobacco (Vineis et al., 1988).

The epidemiological evidence about the carcinogenic risk from coffee drinking, artificial sweeteners, phenacetin, and chlorinated water cannot be considered definitive (refer to Chapters 4 and 5 in this volume).

Data from South Africa, Mozambique, and Egypt consistently indicate an association between *Schistosoma haematobium* and squamous cell carcinoma of the bladder.

Kidney (ICD IX 189)

Renal cancer is not very frequent but has shown a consistent upward trend in most countries. Tobacco smoking is the major known risk factor associated with this tumour. Obesity, coffee, and alcohol drinking, and abuse of phenacetin have been reported in association with elevation of renal cancer risk in some studies, but the evidence is not consistent.

Occupational exposure to metals such as cadmium and lead have been associated to an increased risk of renal cancer.

Brain and Nervous System (ICD IX 191–192)

Neoplasms of the nervous system including brain are difficult to record and to study due to limitations in diagnosis accuracy. Furthermore benign tumours are often clinically as relevant as the malignant ones.

Higher rates for these tumours are present in Jewish population and in Iceland, while low rates are found in Far East countries; Europe and the U.S. occupy intermediate positions.

Increasing time trends both in incidence and mortality have been described in many countries (Davis et al., 1990), which could partly be due to improved diagnostic ability.

Little is known about the etiology of these tumours, which has been associated with occupational exposures to electromagnetic fields, the petrochemical industry, and to agrochemicals among farmers.

An excess of brain tumours has been reported among children with parental exposure to solvents and other chemicals but the epidemiological evidence is not fully consistent. An association between prenatal X-ray indication and childhood brain tumours has also been reported.

Hodgkin's Disease (ICD IX 201)

This disease is not very frequent and is more common in North America and in the northern part of Italy. Low risk areas appear to be Japan and China. Incidence rates are reported rather stable over time, while mortality has been decreasing due to the improved survival (Rosenberg, 1989).

Several episodes of space-time clustering have suggested an infectious origin of the disease. However, the epidemiological evidence is not consistent (Grufferman and Delzell, 1984). Epstein-Barr virus has been suggested (Mueller et al., 1989; Herbst et al., 1990) to be involved in the etiopathogenesis of Hodgkin's disease (HD), although other viruses might be involved as well as genetic factors.

Excesses of HD have been found in teachers, physicians, and nurses, while occupational exposure to solvents, phenoxy acids, and chlorophenols has been reported to increase the risk of HD (Hardell and Bengtsson, 1983).

Non-Hodgkin's Lymphomas (ICD IX 200, 202)

The rates for this group of lymphomas are generally higher than for HD. The geographical distribution shows higher rates in North America, Europe (e.g., Nordic countries and Africa) and lower rates in Asia and South America.

Both incidence and mortality are increasing in most countries, more · sharply in low risk areas. Changes in diagnostic criteria have certainly influenced this pattern which is, however, believed to be real (Barnes et al., 1986).

Burkitt's lymphoma (ICD IX 200) is frequent in tropical Africa and in New Guinea, while rather rare outside this area, although lymphomas like Burkitt's are diagnosed in Western countries more frequently than in the past. There is consensus about the etiologic role of Epstein-Barr virus together with triggering action of malaria infection in Africa (Day et al., 1985; Geser et al., 1989).

Excesses of non-Hodgkin's lymphoma (NHL) have been detected by several studies of forestry and agricultural workers exposed to phenoxy acids, chlorophenols, pesticides, and herbicides (Hardell and Axelson, 1986; Hoar et al., 1986; Woods et al., 1987; Reif et al., 1989).

Immunodeficiency has often been associated with NHL and this link has been dramatically confirmed in AIDS patients.

Multiple Myeloma (ICD IX 203)

This malignant proliferation of plasma cells mainly affects older people (age 75 or more). Both incidence and mortality are increasing for this disease more rapidly in low risk areas (Cuzick et al., 1983). This is due, at least in part, to the improved diagnostic ability.

The etiology of this disease is basically unknown. Increased risks have been associated with occupational exposure to agents in the rubber and the petroleum industry and in farming; and to asbestos, heavy alkali, and carbon monoxide (Linet et al., 1987; Williams et al., 1989; Boffetta et al., 1989) but the results are not fully consistent with other studies.

Leukaemias (ICD IX 204–209)

This group of neoplasms constitutes less than 5% of all cancers and does not present geographical variations as large as for other cancers. Higher rates affect, however, populations of North America, Western Europe, and Israel, while lower rates are registered in Africa and Asia. Incidence and mortality rates have been rather stable over time in adults, while mortality has been decreasing since the sixties in children due to improved efficacy of therapy. Leukaemia, ALL (Acute Lymphatic Leukemia) in particular, is the most frequent neoplasm affecting children.

Ionising radiation is the most important recognised cause of leukaemia. Epidemiological evidence comes from several studies including those on atomic bomb survivors, and other populations exposed to atomic bomb tests (Caldwell et al., 1980). Recently excess risk has been reported among children possibly directly exposed, or through the father, to radiation from nuclear power plants (Gardner et al., 1990a and b).

There is definitive evidence of an increased risk of leukaemia in cases of HD and of ovarian cancer treated with anti-neoplastic drugs (Kaldor et al., 1990a and b).

Benzene has been demonstrated by occupational studies to be a cause of leukaemia, while several other occupational and environmental exposures, like solvents, electromagnetic fields, some pesticides and herbicides, TCE (tetra-chlore ethylene), etc., have been related to increased risk of leukaemia. Difficulties in ascertaining exposure so far do not allow firm conclusions.

Viruses have been long suspected to cause leukaemia, but the epidemiological evidence is still weak. Some retroviruses have been recently associated with T-cell leukaemia.

Environmental Carcinogens

It appears evident from this summary review that the great majority of factors involved in the carcinogenic process are of exogenous origin, generally defined as "environmental".

This document is, however, limited to those factors concerning the general environment in which the population at risk is located. (Personal dietary habits and sexual and reproductive behaviour are excluded.) The summary Tables 1 and 2 present the available epidemiological knowledge on agents, industrial processes, and other exposures of potential environmental interest subdivided into established carcinogens (Table 1) and suspected carcinogens (Table 2).

Several established or suspected carcinogens investigated in the occupational environment have been included on the assumption that they may represent a potential risk for the environment through air and water pollution, wastes, and accidents. Also infectious agents have been included when related to endemic situations.

Lung cancer, which represents the most frequent target organ, has been reviewed elsewhere (refer to Chapter 10 in this volume).

Table 1 includes 20 agents or industrial processes, the majority of which have been investigated through occupational studies. Some of these industrial processes do not exist anymore in several industrialized countries, but are still present in developing countries. The target organs most frequently represented are bladder, skin, and liver.

Table 2 includes 23 agents, industrial processes and exposures suspected to increase the risk of cancer other than lung in humans. As for Table 1, most of the epidemiological evidence derives from occupational studies. The lym-

**Table 1   Agents, Industrial Processes, and Other Exposures Which Imply Potential
Environmental Exposure Carcinogenic to Humans**

Agent/Industrial Process or Exposure	Main Potential Sources of Exposure	Main Organs on Which[a] a Carcinogenic Effect Has Been Demonstrated
Aflatoxin	Environmental	Liver
Aluminum production	Industrial	Bladder
4-Aminobiphenyl	Industrial	Bladder
Arsenic and arsenic compounds	Industrial, agricultural, environmental	Skin, liver
Asbestos	Industrial, environmental, domestic	Pleural and peritoneal mesothelioma
Auramine, manufacture of	Industrial	Bladder
Benzene	Industrial	Leukaemia
Benzidine	Industrial	Bladder
Coal-tar pitches/coal tars	Industrial, urban	Skin, larynx, oral cavity, bladder
Coke production	Industrial	Skin, bladder
Epstein-Barr virus	Environmental	Nasopharyngeal carcinoma, Burkitt's lymphoma[b]
Hepatitis B virus	Environmental	Liver
Ionising radiation	War use, environmental, industrial accident	Leukaemia, breast, thyroid, stomach, colon, bladder, multiple myeloma, oesophagus, ovary
Magenta, manufacture of	Industrial	Bladder
2-Naphthylamine	Industrial	Bladder
Nickel and nickel compounds	Industrial	Nose
Schistosoma haematobium	Environmental	Bladder
Soot	Urban	Skin
U.V. radiation	Environmental, life style	Skin (melanoma)
Vinyl chloride	Industrial	Liver angiosarcoma

[a] Lung cancer excluded.
[b] *Plasmodium falciparum* involved as cofactor.

phatic and haematopoietic system is most frequently involved by the studies investigating this group of suspected carcinogens.

It is interesting to notice the frequent inclusion of agriculture as a potential source of exposure to suspected carcinogens.

RESEARCH NEEDS

Role of Vital Statistics or Available Disease-Oriented Databases in Hypothesis Generating or Testing

Epidemiological research on cancer has been using vital statistics and cancer registries for several decades. In most countries, the predominant use has been monitoring of geographical and temporal variations of the different

Table 2   Agents, Industrial Processes, and Other Exposures Which Imply Potential Environmental Exposure Suspected to Be Carcinogenic to Humans

Agent/Industrial Process or Exposure	Main Potential Sources of Exposure	Main Organs on Which[a] Carcinogenic Effect Has Been Suggested
Atrazine	Agricultural, environmental	Ovary, Non-Hodgkin's lymphomas
Auramine (technical-grade)	Industrial	Bladder
Bitumens, extracts of steam-refined and air-refined bitumens	Industrial, urban	Mouth, larynx
1,3-Butadiene	Industrial	Lymphatic and haematopoietic system
Carbon-black extracts	Industrial	Skin, bladder
Carbon tetrachloride	Industrial, agricultural	Liver
Chlordane	Industrial, agricultural, domestic	Lymphatic and haematopoietic system
Chlorophenols/ chlorophenoxy herbicides	Industrial, agricultural, war use	Soft tissue sarcomas, lymphatic and haematopoietic system
Cobalt and cobalt compounds	Environmental, medical	Lymphatic and haematopoietic system
Creosote	Agricultural	Skin
DDT	Industrial, agricultural	Lymphatic and haematopoietic system
Dichlorvos	Environmental, domestic	Leukaemia
Diesel engine exhaust	Environmental, Urban	Bladder, lung
Ethylene dibromide	Industrial, agricultural	Lymphatic and haematopoietic system
Ethylene oxide	Industrial, agricultural	Lymphatic and haematopoietic system
Heptachlor	Industrial, agricultural, domestic	Lymphatic and haematopoietic system
Lead and lead compounds, inorganic	Industrial, urban	Digestive system, kidney
Formaldehyde	Industrial, agricultural, domestic	Lymphatic and haematopoietic system, nose
ELF (extremely low frequency) radiation	Industrial, environmental	Leukaemia, brain
Petroleum refining	Industrial	Skin cancer, leukaemia
Polychlorinated biphenyls	Industrial, agricultural	Skin (melanoma), liver
Styrene and styrene oxide	Industrial	Lymphatic and haematopoietic system
2,3,7,8-Tetrachlorodibenzo-para-dioxin (TCDD)	Industrial, agricultural, war	Soft tissue sarcoma, lymphatic and haematopoietic system

[a] Lung cancer excluded.

cancer sites, in order to infer from these data hypotheses to be further tested through analytical studies.

In countries such as Sweden, Denmark, and Canada record linkage of environmental and occupational databases with cancer registry files has provided prompt results suitable for both hypotheses generating and testing. The quality of information on exposure is, however, rather limited and can easily generate spurious associations.

Progressively, since the beginning of the eighties, computerisation has largely been introduced in hospitals and pathology departments. These data are tendentially more precise for the disease definition than those based on death certificates. It is essential for the epidemiological research on cancer to have access to these databases at nominal level. Furthermore, an effort should be made to harmonize as far as possible the different databases, in order to allow an efficient and easy linkage between them. This would certainly improve the availability of data for epidemiological research.

Large databases on environmental pollutants (e.g., urban air pollution, indoor radon levels) were and are being implemented in many countries. The use of these data has been so far rather limited, often due to the lack of coordination between environmental hygiene activity and epidemiological research. Databases on environmental pollutants should be collected, keeping in mind their possible utilisation by epidemiological research, and priority should be given as far as possible to the exploitation of existing data rather than to expensive project of new databases.

Definition of Disease Study Entity

It is well known that there is a large variability in the quality of the diagnosis of cancer depending on the source of information. Comparison between diagnoses performed at autopsy and those from other sources of information has clearly confirmed and quantified the problem (IARC, 1991).

The progress in diagnostic techniques and in understanding the biology of cancer has modified several disease entities in relative terms and into a more precise classification of the behaviour (i.e., malignant vs. benign).

There is a need of epidemiological research on cancer sites suspected to share common risk factors and on distinct subtypes of neoplasms previously considered as a unique disease or group of diseases (e.g., leukaemias and lymphomas).

Diagnoses of rare tumours are more reliable than in the past and they would provide an interesting subject for epidemiological research on possible environmental risk factors. Collaborative studies on rare tumours should be encouraged.

Individual Susceptibility

With the development of molecular biology and the improved understanding of the molecular mechanism involved in the development of the malignant cell, it is now possible to plan epidemiological studies on the role of genetic and acquired host susceptibility in increasing the risk in humans exposed to environmental carcinogens.

This aspect of research is certainly important when planning environmental studies, although the present knowledge on environmental carcinogens is very limited and the problem of confounding from other unknown risk factors seems to be more relevant.

Exposure Assessment

There is a need of improving exposure assessment in epidemiological studies investigating environmental carcinogens. This information, when existing, is often insufficient and of poor quality.

A number of biological markers of exposure have been made available, and will presumably increase in the future. This achievement can enormously improve the assessment of exposure at the individual level in epidemiological studies.

Hypothesis to Be Tested as Priorities

Epidemiological studies of cancer need a long preparation and several (three to five) years before obtaining results. It is therefore essential to plan the studies following a prioritising process, which should take into account valid and consistent criteria.

In 1986 in Lyon, France, the Occupational Cancer Programme of the IARC convened a group of experts with the purpose of discussing and suggesting a list of criteria for deciding priorities in occupational cancer epidemiology (IARC/CEC, 1986).

This document, although prepared for occupational research, could at least in part contribute to environmental epidemiology with which it shares many aspects.

It should be clear that in both situations the main goal is to identify exposures which, once controlled, would have the maximum impact in reducing the burden of various types of cancer.

If we assume at least some of the criteria proposed in the IARC document, priority should be given to exposures to established or suspected carcinogens occurring at relatively high levels in large populations, thus resulting in a relevant number of attributable cases.

Considering that exposures at the environmental level, with exception of accidents, are generally low, the size of the population exposed appears to be crucial.

Considering these criteria and the content of Tables 1 and 2, priority should be given to epidemiological investigations on populations resident in industrialised areas where carcinogens were produced or used.

Another priority could be justified by frequent report of cancer risk apparently related to exposure to herbicides and pesticides, which is of concern in large populations of most countries.

A further priority relates to the ascertainment of the carcinogenic risk in populations exposed to biological agents, in view of the possible control of the cancer risk through vaccination.

Epidemiological studies of rare tumours could prove to be a valid tool for the detection of unknown carcinogens.

The use for epidemiological purposes of existing databases should be encouraged and implemented.

Cancer, as well known, is the endpoint of a biological process lasting several years, in most cases more than two decades. Studies of populations concerned by recent exposures to known or suspected carcinogens should be discouraged unless a surveillance system for further follow-up is implemented.

REFERENCES

1. Bang, K. M., White, J. E., Ganse, B. L., and Leffall, L. D., Evaluation of recent trends in cancer mortality and incidence among blacks, *Cancer*, 61, 1255-1261, 1988.

2. Barnes, N., Cartwright, R. A., O'Brien, C., Richards, I. D. G., Roberts, B., and Bird, C. C., Rising incidence of lymphoid malignancies-true or false, *Br. J. Cancer*, 53, 393-398, 1986.

3. Béral, V., Evans, H., Shaw, H., and Milton, G., Malignant melanoma and exposure to fluorescent lighting at work, *Lancet*, ii, 290-293, 1982.

4. Blot, W. J. and Fraumeni, J. F., Trends in esophageal cancer mortality among U.S. blacks and whites, *Am. J. Public Health*, 77, 296-298, 1987.

5. Bodmer, W. F., Bailey, C. J., Bodmer, J., Bussey, H. J., Ellis, A., Gorman, P., Lucibello, F. C., Murday, V. A., Rider, S. H., Scambler, P., Sheer, D., Solomon, E., and Spurr, N. K., Localization of the gene for familial adenomatous polyposis on chromosome 5, *Nature*, 328, 614-616, 1987.

6. Boffetta, P., Stellman, S. D., and Garfinkel, S., A case-control study of multiple myeloma nested in the American Cancer Society prospective study, *Int. J. Cancer*, 43, 554-559, 1989.

7. Boyle, P., Hsieh, C. C., Maisonneuve, P., La Vecchia, C., Macyarlone, G. J., Walker, A. M., and Trichopoulos, D., Epidemiology of pancreas cancer, *Int. J. Pancreatol.*, 5, 327-346, 1989.

8. Breslow, A., Prognosis in cutaneous melanoma: Tumour thickness as a guide treatment, in *Pathology Annual*, Part I, Vol. 15, Sommers S. C., and Rosen O. O., Eds., Appleton-Century-Crofts, New York, 1980, 1-22.

9. Brown, L. M., Mason, T. J., Pickle, L. W., Stewart, P. A., Buffler, P. A., Burau, K., Ziegler, R. G., and Fraumeni, J. F. Jr., Occupational risk factors for laryngeal cancer on the Texas Gulf Coast, *Cancer Res.*, 48, 1960-1964, 1988.

10. Buiatti, E., Palli, D., Decarli, A., Amadori, D., Avellini, C., Bianchi, S., Biserni, R., Cipriani, F., Cocco, P., Giacosa, A., Marubini, E., Puntoni, R., Vindigni, C., Fraumeni, J., and Blot, W., A case-control study of gastric cancer and diet in Italy, *Int. J. Cancer*, 44, 611-616, 1989.

11. Caldwell, G. G., Kelley, D. B., and Heath, C. W. Jr., Leukemia among participants in military maneuvers at a nuclear bomb test (Smoky), *JAMA*, 244, 1575-1578, 1980.
12. Cannon-Albright, L. A., Skolnick, M. H., Bishop, T., Lee, R. G., and Burt, R. W., Common inheritance of susceptibility to colonic adenomatous polyps and associated colorectal cancers, *N. Engl. J. Med.*, 319, 533-537, 1988.
13. Cuzick, J., Velez, R., and Doll, R., International variations and temporal trends in mortality from multiple myeloma, *Int. J. Cancer*, 32, 13-19, 1983.
14. Cuzick, J. and Babiker, A. G., Pancreatic cancer, alcohol, diabetes mellitus and gall-bladder disease, *Int. J. Cancer*, 43, 415-421, 1989.
15. Day, N. E. and Muñoz, N., Esophagus, in *Cancer Epidemiology and Prevention*, Schottenfield, D., and Fraumeni, J. F., Eds., Philadelphia, 1982, 526-623.
16. Day, N. E., Smith, P. G., and Lachet, B., The latent period of Burkitt's lymphoma: the evidence from epidemiological clustering, in *Burkitt's Lymphoma: a Human Cancer Model*, IARC Scientific Publications, 60, Lenoir, G. M., O'Conor, G., and Olweny, C. L. M., Eds., International Agency for Research on Cancer, Lyon, France, 1985, 187-195.
17. Davis, D. L., Hoel, D., Fox, J., and Lopez, A., International trends in cancer mortality in France, West Germany, Italy, Japan, England and Wales, and the USA, *Lancet*, 336, 474-481,1990.
18. De Waard, F., Cornelis, J. P., Aoki, K., and Yoshida, M., Breast cancer incidence according to weight and height in two cities of the Netherlands and in Aichi Prefecture, Japan, *Cancer*, 40, 1269-1275, 1977.
19. De Waard, F. and Trichopoulos, D., A unifying concept of the aetiology of breast cancer, *Int. J. Cancer*, 59, 119-125, 1988.
20. Elwood, J. M., Gallagher, R. P., Worth, A. J., Wood, W. S., and Pearson, J. C., Etiological differences between subtypes of cutaneous malignant melanoma: Western Canada melanoma study, *J. Natl. Cancer Inst.*, 78, 37-44, 1987.
21. Fearon, E. R., Cho, K. R., Nigro, J. M., Kern, S. E., Simons, J. W., Ruppert, J. M., Hamilton, S. R., Preisinger, A. C., Thomas, G., Kinzler, K. W., and Vogelstein, B., Identification of a chromosome 18q gene that is altered in colorectal cancers, *Science*, 247, 49-56, 1990.
22. Gardner, M. J., Snee, M. P., Hall, A. J., Powell, C. A., Downes, S., and Terrel, J. D., Results of case-control study of leukemia and lymphoma among young people near Sellafield nuclear plant in West Cumbria, *Br. J. Med.*, 300, 423-429, 1990.
23. Gardner, M. J., Hall, A. J., Snee, M. P., Downes, S., Powell, C. A., and Terrel, J. D., Methods and basic data of case-control study of leukemia and lymphoma among young people near Sellafield nuclear plant in West Cumbria, *Br. J. Med.*, 30, 429-434, 1990.
24. Geser, A., Brubaker, G., and Draper, C. C., Effect of a malaria suppression program on the incidence of African Burkitt's lymphoma, *Am. J. Epidemiol.*, 129, 740-752, 1989.
25. Graham, S., Haughey, B., Marshall, J., Brasure, J., Zielezny, M., Freudenheim, J., West, D., Nolan, J., and Wilkinson, G., Diet in the epidemiology of gastric cancer, *Nutr. Cancer*, 13, 19-34, 1990.
26. Grufferman, S. and Delzell, E., Epidemiology of Hodgkin's disease, *Epidemiol. Rev.*, 6, 76-106, 1984.

27. Hakulinen, T., Andersen, A. A., Malker, B., Pukkala, E., Schon, G., and Tulinius, H., Trends in cancer incidence in Nordic countries, *Acta Pathol. Microbiol. Immunol. Scand.* Sect. A, 94 (Suppl. 288), 62-63, 1986.

28. Hardell, L. and Bengtsson, N. O., Epidemiological study of socioeconomic factors and clinical findings in Hodgkin's disease, and reanalysis of previous data regarding chemical exposure, *Br. J. Cancer*, 48, 217-225,1983.

29. Hardell, L. and Axelson, O., Phenoxyherbicides and other pesticides in the etiology of cancer: some comments on the Swedish experiences, in *Cancer Prevention, Strategies in the Workplace,* Becker, C. E., and Coye, M. J., Eds., Hemisphere, Washington, D.C., 1986, 107-119.

30. Henderson, B. E., Ross, R. K., and Bernstein, L., Estrogensas a cause of human cancer, The Richard and Hinda Rosenthal Foundation Award Lecture, *Cancer Res.*, 48, 246- 253, 1988.

31. Herbst, H., Niedobitek, G., Kneba, M., Hummel, M., Finn, T., Anagnostopoulos, I., Bergholz, M., Krieger, G., and Stein, H., High incidence of Epstein-Barr virus genomes in Hodgkin's disease, *Am. J. Pathol.*, 137, 13-18, 1990.

32. Higginson, J., Muir, C. S., and Muñoz, N., Eds., Cambridge Monographs on Cancer Research, *Human Cancer: Epidemiology and Environmental Causes*, Cambridge University Press, Cambridge, 1992.

33. Hill, P., Wynder, E. L., Garbaczewski, L., Garnes, H., and Walker, A. R., Response to luteinizing releasing hormone, thyrotrophic releasing hormone, and human chorionic gonadotrophin administration in healthy men at different risks for prostatic cancer and in prostatic cancer patients, *Cancer Res.*, 42, 2074-2080, 1982.

34. Hirayama, T., Actions suggested by gastric cancer epidemiological studies in Japan, in *Gastric Carcinogenesis,* Reed, P. I., and Hill, M. J., Eds., Elsevier Science Publishers, Amsterdam, 1988, 209-277.

35. Hoar, S. K., Blair, A., Holmes, F. F., Boysen, C. D., Robel, R. J., Hoover, R., and Fraumeni, J. F. Jr., Agricultural herbicide use and risk of lymphoma and soft-tissues arcoma, *JAMA*, 256, 1141-1147, 1986.

36. IARC/CEC Working Group Report, *Priorities in Occupational Cancer Epidemiology*, IARC Internal Technical Report No. 86/004, International Agency for Research on Cancer, Lyon, France, 1986.

37. IARC/WHO, *Cancer: Causes, Occurrence and Control*, IARC Scientific Publications No. 100, Tomatis, L., Lyon, France, 1990.

38. IARC, *Autopsy in Epidemiology and Medical Scientifc Research*, IARC Scientific Publications No. 112, Riboli, E., and Delendi, M., Eds., Lyon, France, 1991.

39. IARC, *The Epidemiology of Cervical cancer and Human papillomavirus*, IARC Scientific Publications No. 119, Muñoz, N., Bosch, F. X., Shah, K. V., and Meheus, A., Eds., Lyon, France, 1992.

40. IARC, *Solar and Ultraviolet Radiation,* Vol. 55, IARC Monographs on the Evaluation of Carcinogenic Risks to Humans, Lyon, France, 1992.

41. Jayant, K. and Yeole, B. B., Cancer of the upper alimentary and respiratory tract in Bombay, India: a study of incidence over two decades, *Br. J. Cancer*, 56, 847-852, 1987.

42. Kaldor, J. M., Day, N. E., Pettersson, F., Clarke, E., Pedersen, D., Mehnert, W., Bell, J., Host, H., Prior, P., Karjalainen, S., Neal, F., Koch, M., Band, P., Choi, W., Pompe Kirn, V., Arslan, A., Zarén, B., Belch, A. R., Storm, H., et al., Leukemia following chemotherapy for ovarian cancer, *N. Engl. J. Med.*, 322, 1-6, 1990.

43. Kaldor, J. M., Day, N. E., Clarke, E., Van Leeywen, F. E., Henry-Amar, M., Fiorentino, M. V., Bell, J., Pedersen, D., Band, P., Assouline, D., Koch, M., Choi, W., Prior, P., Blair, V., Langmark, F., Pompe Kirn, V., Neal, F., Peters, D., Pfeiffer, R., Karjalainen, S., et al., Leukemia following Hodgkin's disease, *N. Engl. J. Med.*, 322, 7-13, 1990.

44. La Vecchia, C., Negri, E., Decarli, A., D'Avanzo, B., and Franceschi S., A case-control of diet and gastric cancer in Northern Italy, *Int. J. Cancer*, 40, 484-498, 1987.

45. Linet, M. S., Harlow, S. D., and McLaughlin, J. K., A case-control study of multiple myeloma in whites: chronic antigenic stimulation, occupation, and drug use, *Cancer Res.*, 47, 2978-2981, 1987.

46. Mueller, N., Evans, A., Harris, N. L., Comstock, E., Jellum, E., Magnus, K., Orentreich, N., Polk, F., and Vogelman, J., Hodgkin's disease and Epstein-Barr virus, Altered antibody pattern before diagnosis, *N. Engl. J. Med.*, 60, 321-332, 1989.

47. Muñoz, N. and Bosch, F. X., Epidemiology of hepatocellular carcinoma, in *Neoplasms of the Liver,* Okuda, K., and Purchase, I. F., Springer, Tokyo, 1987, 3-19.

48. Muñoz, N., Descriptive epidemiology of stomach cancer, in *Gastric Carcinogenesis*, Reed, P. I., and Hill, M. J., Eds., Elsevier Science Publishers, Amsterdam, 1988, 51-69.

49. Muñoz, N. and Bosch, F. X., Epidemiology of cervical cancer, in *Human papillomavirus and Cervical Cancer,* Muñoz, N., Bosch, F. X., and Jensen, O. M., Eds., IARC Scientific Publications No. 94, International Agency for Research on Cancer, Lyon, France, 1989, 9-39.

50. Osterlind, A., Tucker, M. A., Hou-Jensen, K., Stone, B. J., Englom, G., and Jensen, O. M. I., The Danish case-control study of cutaneous malignant melanoma, importance of host factors, *Int. J. Cancer*, 42, 200-206, 1988.

51. Osterlind, A., Tucker, M. A., Stone, B. J., and Jensen, O. M. II, The Danish case-control study of cutaneous malignant melanoma, importance of UV light exposure, *Int. J.Cancer*, 42, 319-324, 1988.

52. Parkin, D. M., Läära, E., and Muir, C. S., Estimates of the worldwide frequency of sixteen major cancers in 1980, *Int. J. Cancer*, 41, 184-197, 1988.

53. Reif, J., Pearce, N., Kawachi, I., and Fraser, J., Soft tissue sarcoma, non-Hodgkin's lymphoma and other cancer in New Zealand forestry workers, *Int. J. Cancer*, 43, 49-54, 1989.

54. Rosenberg, S. A., Hodgkin's disease: challenges for the future, *Cancer Res.*, 49, 767-769, 1989.

55. Ross, R. K., Smimizu, H., Paganini-Hill, A., Honda, G., and Henderson, B. E., Case-control study of prostate cancer in blacks and whites in Southern California, *J. Natl. Cancer Inst.*, 78, 869-874, 1987.

56. Tuyns, A. J., Incidence trends of laryngeal cancer in relation to national alcohol and tobacco consumption, in *Trends in Cancer Incidence — Causes and Implications*, Magnus K., Ed., Hemisphere, New York, 1982, 199-214.
57. Tuyns, A. J., Estève, J., Raymond, L., Berrino, F., Benhamou, E., Blanchet, F., Boffetta, P., Crosignani, P., Del Moral, A., Lehmann, W., Merletti, F., Péquinot, R., Riboli, E., Sancho-Garnier, H., Terracini, B., Zubiri, A., and Zubiri, L., Cancer of the larynx/hypopharynx, tobacco and alcohol: IARC international case-control study in Turin and Varese (Italy), Zaragoza and Navarra (Spain), Geneva(Switzerland) and Calvados (France), *Int. J. Cancer*, 41, 483-491, 1988.
58. Victora, C. G., Muñoz, N., Day, N. E., Barcelos, L. B., Peccin, D. A., and Braga, N. M., Hot beverages and oesophageal cancer in Southern Brazil: a case-control study, *Int. J. Cancer*, 39, 710-716, 1987.
59. Vineis, P., Estève, J., Hartge, P., Hoover, R., Silverman, D.T., and Terracini, B., Effects of timing and type of tobacco in cigarette-induced bladder cancer, *Cancer Res.*, 48, 3849-3852, 1988.
60. Walter, S. D., Marrett, D., From, L., Hertzman, C., Shannon, H. S., and Roy, P., The association of cutaneous malignant melanoma with the use of sunbeds and sunlamps, *Am. J. Epidemiol.*, 131, 232-240, 1990.
61. Williams, A. R., Weiss, N. S., Koepsell, T. D., Lyon, J. L., and Swanson, G. M., Infectious and noninfectious exposures in the etiology of light chain myeloma: a case-control study, *Cancer Res.*, 49, 4038-4041, 1989.
62. Woods, J. S., Polissar, L., Severson, R. K., Heuser, L. S., and Kulander, B. G., Soft tissue sarcoma and non-Hodgkin's lymphoma in relation to phenoxy herbicide and chlorinated phenol exposure in western Washington, *J. Natl. Cancer Inst.* 78, 899-890, 1987.
63. Zaridze, D. G. and Boyle, P., Cancer of the prostate: epidemiology and aetiology, *Br. J. Urol.*, 59, 493-502, 1987.

10   LUNG CANCER

Göran Pershagen

CONTENTS

INTRODUCTION

Lung cancer is believed to be the most common fatal neoplastic disease in the world today (IARC, 1986). In comparison with other types of cancer, many etiological factors have been identified, which together often account for a major part of the cases. Tobacco smoking is the dominating cause and occupational exposures are of importance in some situations. However, for many environmental exposures, which may be of great significance from a public health point of view, the evidence remains unclear.

This review will focus on lung cancer in relation to environmental exposures such as ambient air pollution, residential radon, environmental tobacco
smoke (ETS), and combustion products from indoor sources. Tobacco smoking
and occupational exposures are taken up mainly as interacting and confounding
factors and for risk extrapolation from high exposures. Identification of
research needs will receive particular attention, especially in relation to the
application of new molecular biologic techniques in epidemiologic studies.

LUNG CANCER OCCURRENCE

The incidence of lung cancer has risen sharply during this century, and it
is now one of the leading causes of cancer death in many countries. For
example, in the U.S. the age-adjusted lung cancer mortality in men increased
from 11 per 100,000 in 1940 to 73 per 100,000 in 1982 (Garfinkel & Silverberg, 1990). The lung cancer death rate in women started to rise in the early
1960s from 6 per 100,000 to 25 per 100,000 in 1986. In the younger age
groups there has been a levelling off in the mortality rates among men but not
in women. In Sweden the lung cancer incidence has increased an average of
1.7 and 3.9% yearly in men and women, respectively, during the last two
decades (Pershagen, 1990). Lung cancer rates among middle-aged men in
Eastern Europe are approaching the high rates in Finland and the U.K. of the
1950s (IARC, 1986). Also in many developing countries there is a sharp
increase in lung cancer incidence and by the turn of the century it is estimated
that around 2 million lung cancer deaths will appear annually in the world.

There are pronounced differences in lung cancer incidence between different countries and regions. The highest age standardised incidence rates in
an international comparison were found in black men of New Orleans, U.S.
(110 per 100,000 and year) and in Maori women of New Zealand (68 per
100,000 and year). The lowest rates in men and women were found in Madras,
India (6 and 2 per 100,000 and year, respectively). In general, urban areas
showed higher rates in both men and women than rural areas (IARC, 1987).

The major differences in lung cancer occurrence between high and low
incidence areas indicate that there is a substantial potential for prevention.
Furthermore, in spite of great efforts to improve treatment, most lung cancer
patients die within 1 year after diagnosis. Primary prevention must thus be
given high priority.

ENVIRONMENTAL CAUSES OF LUNG CANCER

Tobacco Smoking and Occupation

Tobacco smoking is the most important cause of lung cancer, and where
prolonged smoking is widespread it generally accounts for more than 80% of

the cases (IARC 1986). The risk is closely related to duration and intensity of smoking. After cessation of smoking the excess relative risk of lung cancer falls and will approach zero after 10–20 years (Cederlöf et al., 1975; Doll & Peto, 1976). Studies from the U.K. and the U.S. have shown lower risks for lung cancer in pipe smokers than in cigarette smokers (Doll & Peto, 1976, Rogot & Murray, 1980), but this is not true for Sweden (Damber & Larsson, 1987; Carstensen et al., 1987). The dominating role of smoking in lung cancer makes it important to be considered as an effect modifier and/or confounder in all studies of lung cancer etiology.

A number of occupational exposure factors have been associated with an increased risk of lung cancer, such as asbestos, chromates, inorganic arsenic, polyaromatic hydrocarbons, and radon (Simonato et al., 1988). Although individual risks may be substantial, the population attributable proportion of lung cancer due to occupational exposures is generally low because of the limited number of subjects exposed. However, in some situations where the occupationally exposed constitute a large fraction of the population, the attributable proportion may reach 40%.

In several studies on lung cancer etiology, a multiplicative interaction between smoking and occupational exposures has been observed. Some examples include asbestos (Hammond et al., 1979), arsenic (Pershagen et al., 1981), and radon daughters (Archer et al., 1973). Multistage models for carcinogenesis with agents operating at different stages in the cancer induction process may generate this type of interaction. In other studies, the interaction has been less pronounced, more consistent with an additive effect (Pinto et al., 1978, Radford & St. Clair-Renard, 1984). In most studies it was not possible to conclusively reject either an additive or a multiplicative model, mainly because of a small number of non-smoking lung cancer cases resulting in a low statistical power of the test.

Ambient Air Pollution

Urban Areas

The composition of ambient air in urban areas is quite variable and complex. Some examples of the environments under investigation in the epidemiologic studies reviewed here include British towns and cities during the 1950s; and urban areas in Japan, China, Europe, and the U.S. from the 1960s to the 1980s. Emissions resulting from the use of coal and other fossil fuels for residential heating were dominating sources of pollution in some areas, while in others motor vehicles or industries were more important. The term "urban" thus denotes a mixture of environments, which may show substantial differences, both in terms of actual exposures to various environmental pollutants and the influence of interacting or confounding factors.

The epidemiologic evidence on air pollution and lung cancer has recently been reviewed by Pershagen and Simonato (1990) and will be discussed only briefly here. Seven cohort studies on urban air pollution and lung cancer were available. All but one of the investigations contained information on smoking for all study subjects. The studies came from the U.S. (3), Sweden (2), Finland (1), and the U.K. (1). Smoking-adjusted relative risks for lung cancer in urban areas were generally on the order of 1.5 or lower in those cohort studies reporting increased risks. The findings pertain mainly to smokers. For non-smokers the number of cases was generally too small for a meaningful inter-pretation of urban-rural differences.

In the 13 case-control studies reviewed by Pershagen and Simonato (1990), residential and smoking histories were obtained for the study subjects and sometimes also information on potential confounding factors, such as occupation. Increased relative risks for lung cancer were observed among men in urban areas in three British studies as well as in studies from Greece, Poland, China, and Japan. Two U.S. studies found raised lung cancer risks in urban males, while another two failed to show an effect. The results for women are difficult to interpret because of small numbers, but at least one study indicated a raised lung cancer risk for females in urban areas, also among nonsmokers. The magnitude of the excess relative risks for lung cancer in urban areas reported in the case-control studies was similar to that in the cohort studies.

The epidemiologic studies on urban air pollution and lung cancer gave somewhat inconsistent results as to the type of interaction with tobacco smok-ing. Some studies provided evidence of a combined effect exceeding an addi-tive effect, and often compatible with a multiplicative interaction, while other studies were more consistent with an additive effect.

Industrial Areas

Several epidemiologic studies have been carried out in areas near copper, lead, or zinc smelters (Pershagen & Simonato, 1990). The emissions from the smelters are quite complex, but inorganic arsenic is often a major component. The studies come from four countries (Canada, China, Sweden, and U.S.) and are of ecologic or case-control design. Five ecologic studies showed increased lung cancer rates among men living in areas near non-ferrous smelters with relative risks ranging from about 1.2 to over 2. Two case-control studies showed relative risks of 1.6 and 2.0 for men living near the smelters after adjustment for occupation and smoking. A study of women near a U.S. smelter also suggested an increased lung cancer risk related to estimated exposure to arsenic in ambient air.

Ecologic studies on lung cancer have been performed in areas with indus-tries of different types, including chemical, pesticide, petroleum, shipbuilding, steel, and transportation industries (Pershagen & Simonato, 1990). Most of the studies showed increased lung cancer risks, which did not seem to be explained by socioeconomic factors. However, smoking habits were not controlled or was employment at the industries under study.

It is difficult to interpret the epidemiologic evidence on ambient air pollution and lung cancer. Many studies were not originally designed to study this relation which has implications for the detail and quality of the exposure information. Data from measurements of air pollutants were generally limited, making it difficult to compare the findings in different studies and to assess dose-response relationships. Uncontrolled confounding may also have contributed to the results.

Residential Radon

A number of epidemiologic investigations show that underground miners exposed to high levels of radon daughters run an increased risk of lung cancer (NAS, 1988). Studies in experimental animals confirm that inhalation of radon daughters can induce lung cancer. The increased radon concentrations found in many homes of some countries, suggest that residential radon exposure is an important risk factor for lung cancer in the general population. However, for several reasons, quantitative assessments of population risks based on the data in miners and experimental animals are uncertain.

Lung cancer risks related to residential radon exposure have been studied with different epidemiologic methodologies (Samet, 1989). Earlier cohort and case-control studies based the exposure estimation primarily on housing characteristics and/or geology. Many of these studies showed an association between estimated radon exposure and lung cancer risk; however, limited study sizes and imprecision in the exposure estimates make it difficult to use the data for quantitative risk assessments.

Five recent case-control studies based the exposure estimation on radon measurements in homes of the study subjects covering about 10–30 years of residency (Axelson et al., 1988; Blot et al., 1990; Schoenberg et al., 1990; Ruosteenoja, 1991; Pershagen et al., 1992). These studies also contained detailed information on smoking and other potential confounding factors. Positive exposure-response relationships were suggested in some of the studies, while others found no evidence of a trend. Differences between the studies in exposure levels and in the influence of other risk factors complicate a comparison of the results.

The risk estimates in the studies on residential radon exposure and lung cancer may be compared with those obtained from miners. Risk estimations based on the mining data often used relative risk models, either constant (ICRP, 1987) or modified by age and time since exposure (NAS, 1988). The risk estimates from the residential radon studies by Schoenberg et al. (1990), Ruosteenoja (1991), and Pershagen et al. (1992) appear to lie within the same range as those projected from miners, while the study by Blot et al. (1990) suggests a lower risk. There are considerable uncertainties in the comparisons with the mining data. For example, age differences in risk or consequences of a nonmultiplicative interaction between radon and smoking have not been considered.

The type of interaction between radon exposure and smoking is of importance for the risk assessment. In miners a multiplicative or submultiplicative interaction was often found but the data are not fully consistent (NAS, 1988). A major difficulty in the interpretation arises from the fact that the number of lung cancers is low among nonsmokers, although the data clearly indicate that the risk is elevated also in this group among radon exposed miners. For residential exposure only limited evidence is available and no clear pattern of interaction between smoking and radon exposure has emerged.

Cooking Fumes

Several epidemiologic studies have investigated lung cancer risks in relation to exposure to combustion products in homes. The sources include fuels used for cooking and heating, such as coal and kerosene, as well as vegetable oils used for frying, etc. Most of the studies come from China, where indoor air pollution levels from such sources may be substantial (Mumford et al., 1987). In this context it is worth noting that increased lung cancer risks have been reported both among cooks and bakers (Coggon et al., 1986; Tüchsen & Nordholm, 1986), and among coke oven workers (Redmond, 1976).

Two case-control studies on female lung cancer from China suggest that cooking oil fumes may be of importance for lung cancer (Gao et al., 1987; Wu-Williams et al., 1990). In one of the studies there was an increased risk of lung cancer associated with the use of rapeseed oil for cooking compared to soybean oil. The lung cancer risk was related to degree of smokiness when cooking and number of dishes prepared per week. In the other study increased risks of lung cancer were related to number of meals prepared by deep frying and to eye irritation during cooking.

In parts of China indoor burning of coal for domestic cooking and heating is common. Two types of coal are "smoky coal", which smokes heavily on firing; and "smokeless coal", which produces little smoke. Lung cancer rates in Xuan Wei county among both men and women living in areas where "smoky" coal was the predominant fuel were substantially higher than in areas where "smokeless" coal or wood was used (Mumford et al., 1987). A subsequent case-control study indicated that domestic coal use was a stronger risk factor for lung cancer in females than smoking and that the risk was related to the duration of exposure (He et al., 1991). In the study by Wu-Williams et al. (1990) there was an elevated risk related to the number of years using coal for domestic heating.

A case-control study from Guangzhou, China showed that several variables indicating a poor kitchen and house ventilation were associated with an increased lung cancer risk (Liu et al., 1992). Coal was the predominant fuel used in this area. An elevated relative risk associated with residential coal burning during childhood was observed in a U.S. case-control study (Wu et al., 1985).

A case-control study from Hong Kong (Koo et al., 1983) indicated an increased lung cancer risk in women associated with kerosene use. No excess risk was observed for other types of fuels, such as wood/grass or liquid petroleum gas. However, in a study of women from Osaka, Japan, an increased lung cancer risk was related to the use of straw or wood as cooking fuel (Sobue et al., 1990).

It may be concluded that indoor combustion sources can be of importance for cancer of the respiratory tract. In some situations, such as in parts of China where smoking is rare and indoor burning of coal is common, it seems to be the dominant cause of lung cancer. It is possible that fumes from vegetable oils used for cooking may also play a role, but these data need confirmation.

Environmental Tobacco Smoke

Environmental tobacco smoke is produced from the side stream smoke (SS) released from the burning tobacco product and mainstream smoke (MS) exhaled by the smoker. There are substantial differences in the composition of mainstream and side stream smoke (NRC, 1986). As a rule the amounts emitted from a cigarette are greater in the side stream smoke. For example, more than 50 times higher amounts are emitted of some nitrosamines in the side stream smoke, but the SS/MS ratio is around 2–5 for most components. ETS is predominantly an indoor problem; and the exposure level depends on the intensity of smoking, room size, and air exchange.

Exposure to ETS can be estimated with biologic markers. Nicotine and its metabolite cotinine are two sensitive and specific markers which may be measured in saliva, serum, and urine. Cotinine has some advantages in comparison with nicotine, including a longer biologic half-life. The levels of cotinine in nonsmokers reporting ETS exposure range from about one half to a few per cent of those in regular smokers (Jarvis et al., 1984; Husgafvel-Pursianen et al., 1987). High urinary cotinine concentrations have been reported in children with smoking parents indicating a substantial exposure to ETS (Greenberg et al., 1984; Henderson et al., 1989; Rylander et al., 1989).

The evidence on ETS and lung cancer has recently been reviewed by Pershagen (1994) and will only be discussed briefly here. A total of almost 3000 non-smoking lung cancer cases have been included in 27 studies, suggesting a total study base on the order of 40 million person-years under observation. Most of the studies were of case-control design, and only 3 were cohort studies. The studies come from six countries (China with Hong Kong, Greece, Japan, Sweden, U.K., and U.S.) and include predominantly women (more than 90% of the cases).

Exposure to ETS was defined in different ways in the studies; however, very often the exposure classification was based on smoking habits of spouses. Combining the results of the studies gives pooled relative risks of lung cancer in nonsmoking women and men living with smokers of 1.23 (1.11–1.36) and 1.82 (0.98–3.37), respectively.

It is necessary to consider various sources of bias and their implications in the assessment of the epidemiologic evidence on passive smoking and lung cancer. This is particularly relevant in view of the weak associations observed. Probably the most important source of bias is confounding by unreported active smoking. There is a tendency for smokers to marry smokers; and in conjunction with some misclassification of smokers as nonsmokers, this would tend to produce a spurious association between spouse smoking and lung cancer. Unfortunately there is limited empirical information which can be used to estimate the extent of this bias, and differences in assumptions may lead to quite discrepant conclusions regarding its magnitude (Wald et al., 1986; Lee, 1988). Other types of bias, such as non differential misclassification of ETS exposure and lung cancer, may give rise to an underestimation of the true risk.

RESEARCH NEEDS

Role of Vital Statistics

As indicated previously cancer and mortality registers have played a significant role in studies of lung cancer occurrence and etiology. Because of the dominating role of tobacco smoking for lung cancer, descriptive data on geographical differences and time trends will mainly reflect earlier and present smoking habits. However, there are many examples where descriptive information on lung cancer occurrence has led to the identification of occupational and environmental hazards (Pershagen & Simonato, 1990). It would be particularly useful to obtain more data on lung cancer occurrence in areas with heavy air pollution, such as in parts of Central and Eastern Europe.

The high lethality of lung cancer implies that information on its occurrence also may be obtained also using mortality data. Furthermore, it is possible to assess the quality of both mortality and cancer morbidity registers by comparing the data on lung cancer. For many reasons it is of great importance to have high quality registries on mortality and cancer morbidity. Besides providing useful descriptive information, they may also serve as a source of cases in analytical studies. Thus, efforts should be made to support existing registries and to create new registries, as well as to continuously assess their quality. International collaboration is vital in these activities.

Definition of Disease

The term lung cancer mainly denotes primary cancer of the bronchus or lung. Generally, the quality of this diagnosis is high. However, in nonsmokers, where this tumour is rare, substantial misclassification may occur. Secondary lung tumours and carcinomas with unknown primary site appeared in about one sixth of reported cases of lung cancer on death certificates in the U.S. (Garfinkel, 1981) and in central health registers in Sweden (Pershagen et al., 1987) among female nonsmokers. Descriptive data on time trends of lung cancer in non-smokers must be interpreted with caution and the validity of the diagnostic information should be carefully assessed in analytic studies of lung cancer in nonsmokers.

Lung carcinomas can be separated into different histological types, and international criteria of classification have been proposed (WHO, 1982). Environmental exposures may be associated with specific histological types. For example, tobacco smoking primarily induces epidermoid and small cell carcinomas, although increased relative risks are also found for other histological types (IARC, 1986). Small cell carcinomas seem to predominate among uranium miners, especially in younger age groups (NAS, 1988). However, there is a great need for further data on associations between environmental exposures and specific histological types of lung cancer, which may increase the resolution power of the studies.

New developments in molecular biology have led to the identification of genes which are involved in cancer induction, such as oncogenes and tumour-suppressor genes. Ki-ras and p53 mutations are common in human lung cancers (Bos, 1990; Hollstein et al., 1991). A recent study of uranium miners found differences from the usual lung cancer mutational spectrum in these two genes, which may reflect the genotoxic effects of radon (Vähäkangas et al., 1992). Further studies should be performed to identify relevant mutations for lung cancer caused by environmental exposures.

Individual Susceptibility

A genetically determined individual susceptibility to lung cancer has been suggested. Variation in the ability to metabolise xenobiotics is considered as a possible explanation. Interest has focused on polymorphism of debrisoquine metabolism, aryl hydrocarbon hydroxylase, and glutathione transferase activity (Idle, 1991). The evidence is not entirely consistent, which may have to do with differences in design and selection of study subjects. There is a need for further studies applying modern epidemiologic methodology to control bias in the assessment of individual susceptibility to lung cancer. If high risk populations could be identified, there would be better possibilities of detecting effects of environmental exposures.

Markers of Exposure

Crude exposure measures is a major problem in analytic epidemiology, which will generally lead to a dilution of any associations if the imprecision is unrelated to the disease under study (nondifferential misclassification). New developments in molecular biology have created possibilities for increasing the precision of the exposure measurements. For example, exposure to geno-toxic agents may be monitored by measurement of DNA-adducts and muta-tions of marker genes, such as the hypoxanthine-guanine-phosphoribosyl-transferase (hprt) gene. Increased levels of DNA-adducts and hprt-mutations have been observed following exposures both in the occupational and general environment (Hemminki et al., 1990a; Hemminki et al., 1990b; Bridges et al., 1991; Tates et al., 1991). Biologic markers of long-term exposures extending over years or decades would be particularly useful for epidemiologic studies.

HYPOTHESES TO BE TESTED

There is a need for further studies on the effects of ambient air pollution by chemical type or species on lung cancer. In particular, very few studies are available on lung cancer risks in areas where motor vehicles constitute the dominating source of pollution. Studies using modern epidemiologic method-ology should also be conducted in heavily polluted areas, such as in parts of Central and Eastern Europe, to generate data for risk identification under different exposure situations.

Current risk estimates based on miners suggest that residential radon exposure is the second most important cause of lung cancer in some countries and regions. However, more data are needed from epidemiologic studies on radon exposure in residences. For the purpose of increasing the present uncer-tainty in the risk estimation, these studies should be sufficiently large and involve radon measurements of residential periods covering several decades.

More than 25 epidemiologic studies have been published on passive smoking and lung cancer. Taking the studies together, there is a statistically significant increase in the lung cancer risk of about 20–30% in nonsmokers married to smokers. It is probable that a part of this increase is explained by confounding by smoking. More empirical information on the degree of under-reporting of smoking by nonsmokers is needed to estimate the influence of this type of bias.

A few epidemiologic studies indicate that indoor combustion sources may be of importance for lung cancer. Since this type of exposure is common in some parts of the world, particularly for women, there is a need for further evidence as a basis for the risk assessment.

New developments in molecular biology should be utilised to a greater extent in environmental epidemiology. For example, such techniques may be used to identify sensitive groups, to increase the precision of the exposure measures, and to enhance the specificity of the effect determinations. The employment of molecular biologic techniques in epidemiology could increase the resolution power of the studies and contribute to elucidating mechanisms of toxic effects.

REFERENCES

Archer, V. E., Wagoner, J. K., and Lundin, F. E., Uranium mining and cigarette smoking effects on man. *J. Occup. Med.*, 15, 204, 1973.

Axelson, O., Andersson, K., Desai, G., et al., Indoor radon exposure and active and passive smoking in relation to the occurrence of lung cancer. *Scand. J. Work Environ. Health*, 14, 286, 1988.

Blot, W. J., Xu, Z. Y., Boice, J. D. Jr., et al., Indoor radon and lung cancer in China. *J. Natl. Cancer Inst.*, 82, 1025, 1990.

Bos, J. L., Ras oncogenes in human cancer: A review. *Cancer Res.*, 49, 4682, 1990.

Bridges, B. A., Cole, J., Arlett, C. F., et al., Possible association between mutant frequency in peripheral lymphocytes and domestic radon concentrations. *Lancet*, 337,1187,1991.

Carstensen, J. M., Pershagen, G., and Eklund, G., Mortality in relation to cigarette and pipe smoking: 16 years' observation of 25000 Swedish men. *J. Epidemiol. Commun. Health*, 41, 166, 1987.

Cederlöf, R., Friberg, L., Hrubec, Z., et al., The relationship of smoking and some social covariables to mortality and cancer morbidity, Department of Environmental Hygiene, The Karolinska Institute, Stockholm, 1975.

Coggon, D., Pannett, B., Osmond, C., et al., A survey of cancer and occupation in young and middle aged men. I. Cancers of the respiratory tract. *Br. J. Ind. Med.*, 43, 332, 1986.

Damber, L. and Larsson, L. G., Lung cancer in males and type of dwelling — An epidemiologic pilot study. *Acta Oncol.*, 26, 211, 1987.

Doll, R. and Peto, R., Mortality in relation to smoking: 20 years' observation on male British doctors. *Br. Med. J.*, 2, 1525, 1976.

Gao, Y. T., Blot, W. J., Zheng, W., et al., Lung cancer among Chinese women. *Int. J. Cancer*, 40, 604, 1987.

Garfinkel, L., Time trends in lung cancer mortality among nonsmokers and a note on passive smoking. *J. Natl. Cancer Inst.*, 66, 1061, 1981.

Garfinkel, L. and Silverberg, E., Lung cancer and smoking trends in the U.S. over the past 25 years. *Ann. N.Y. Acad. Sci.*, 609, 146, 1990.

Greenberg, R. A., Haley, N. J., Etzel, R., et al., Measuring the exposure of infants to tobacco smoke. *N. Engl. J. Med.*, 310, 1075, 1984.

Hammond, E. C., Selikoff, I. J., and Seidman, H., Asbestos exposure, cigarette smoking and death rates. *Ann. N.Y. Acad. Sci.*, 330, 473, 1979.

He, X. Z., Chen, W., Liu, Z., et al., An epidemiological study of lung cancer in Xuan Wei county, China: Current progress. Case-control study on lung cancer and cooking fuel. *Environ. Health Perspect.*, 94, 9, 1991.

Hemminki, K., Randerath, K., Reddy M. V., et al., Postlabeling and immunoassay analysis of polycyclic aromatic hydrocarbon adducts of deoxyribonucleic acid in white blood cells of foundry workers, *Scand. J. Work Environ. Health*, 16, 158, 1990a.

Hemminki, K., Grzybowska, E., Chorazy, M., et al., DNA adducts in humans environmentally exposed to aromatic compounds in an industrial area of Poland. *Carcinogenesis*, 11, 1229, 1990b.

Henderson, F. W., Reid, H. F., Morris, R., et al., Home air nicotine levels and urinary cotinine excretion in preschool children. *Am. Rev. Respir. Dis.*, 140, 197, 1989.

Hollstein, M., Sidransky, D., Vogelstein, B., et al., p53 mutations in human cancers. *Science,* 253, 49, 1991.

Husgafvel-Pursianen, K., Sorsa, M., Engström, K., et al., Passive smoking at work: Biochemical and biological measures of exposure to environmental tobacco smoke. Int. Arch. Occup. Environ. Health, 59, 337, 1987.

IARC., Monographs on the evaluation of the carcinogenic risk of chemicals to humans: Tobacco smoking, International Agency for Research on Cancer, Lyon, France, 1986.

IARC., Cancer incidence in five continents. Volume V, International Agency for Research on Cancer, Lyon, France, 1987.

ICRP., Lung cancer risk from indoor exposures to radon daughters. Report No. 50. *Ann. ICRP,* 17, 1, 1987.

Idle, J. R., Is environmental carcinogenesis modulated by host polymorphism? *Mutat. Res.,* 247, 259, 1991.

Jarvis, M. J. and Russel M. A. H., Measurement and estimation of smoke dosage to nonsmokers from environmental tobacco smoke. *Eur. J. Respir. Dis.* 65, 68, 1984.

Koo, L. C., Lee, N., and Ho, J. H. C., Do cooking fuels pose a risk for lung cancer? A case-control study of women in Hong Kong. *Ecol. Dis.* 2, 255, 1983.

Lee, P. N., *Misclassification of Smoking Habits and Passive Smoking,* Springer, Berlin, 1988.

Liu, Q., Sasco, A. J., Riboli, E. et al., Indoor air pollution and lung cancer in Guang Zhou, Peoples Republic of China. *Am. J. Epidemiol.,* (in press).

Mumford, J. L., He, X. Z., Chapman, R. S., et al., Lung cancer and air pollution in Xuan Wei, China. *Science*, 235, 217, 1987.

NAS., Radon and other internally deposited alpha-emitters. BEIR IV, National Academy Press, Washington, D.C., 1988.

NRC., Environmental tobacco smoke: Measuring exposures and assessing health effects, U.S. National Research Council, National Academy Press, Washington, D.C., 1986.

Pershagen, G., Wall, S., Taube, A., et al., On the interaction between occupational arsenic exposure and smoking and its relationship to lung cancer. *Scand. J. Work Environ. Health,* 7, 302, 1981.

Pershagen, G., Hrubec, Z., and Svensson, C., Passive smoking and lung cancer in Swedish women. *Am. J. Epidemiol.,* 126, 17, 1987.

Pershagen, G. and Simonato, L., Epidemiological evidence on air pollution and cancer, in *Air Pollution and Human Cancer,* Thomatis, L., Ed., Springer-Verlag, Heidelberg, 1990.

Pershagen, G., Causes of lung cancer in Sweden, in Health Morbidity and Mortality by Causes of Death in Europe, Juozylynas, A., Ed., Vilnius, Lithuanian Ministry of Health, 1990.

Pershagen, G., Passive smoking and lung cancer, in *Epidemiology of Lung Cancer,* Samet, J., Ed., Marcel Dekker, New York, 1994.

Pershagen, G., Liang, Z. H., Hrubec, Z., et al., Residential radon exposure and lung cancer in women. *Health Phys.,* 63, 179, 1992.

Pinto, S. S., Henderson, V., and Enterline, P. E., Mortality experience of arsenic exposed workers. *Arch. Environ. Health,* 33, 325, 1978.

Radford, E. P. and St Clair-Renard, K. G., Lung cancer in Swedish iron miners exposed to low doses of radon daughters. *N. Engl. J. Med.,* 310, 1485, 1984.

Redmond, C. K., Epidemiological studies of cancer mortality in coke plant workers, in Seventh Conference on Environmental Toxicology 1976, U.S. Environmental Protection Agency, Washington D.C., 1976.

Rogot, E. and Murray, J. L., Smoking and causes of death among U.S. veterans; 16 years of observation. *Public Health Rep.*, 95, 213, 1980.

Ruosteenoja, E., Indoor radon and risk of lung cancer: An epidemiological study in Finland, Finnish Centre for Radiation and Nuclear Safety, Helsinki, 1991.

Rylander, E., Pershagen, G., Curvall, M., et al., Exposure to environmental tobacco smoke and urinary excretion of cotinine and nicotine in children. *Acta Paediatr. Scand.*, 78, 449, 1989.

Samet, J. M., Radon and lung cancer. *J. Natl. Cancer Inst.*, 81, 745, 1989.

Schoenberg, J. B., Klotz, J. B., Wilcox, H. B., et al., Case-control study of residential radon and lung cancer among New Jersey women. *Cancer Res.*, 50, 6520, 1990.

Simonato, L., Vineis, P., and Fletcher A. C., Estimates of the proportion of lung cancer attributable to occupational exposure. *Carcinogenesis*, 9, 1159, 1988.

Sobue, T. R., Suzuki, R., Nakayama, N., et al., Passive smoking among nonsmoking women and the relationship between indoor air pollution and lung cancer incidence. Results of a multicenter case-control study. *Gan No Rinsho*, 36, 329, 1990.

Tates, A. D., Grummt, T., Törnqvist, M., et al., Biological and chemical monitoring of occupational exposure to ethylene oxide. *Mutat. Res.*, 250, 483, 1991.

Tüchsen, F. and Nordholm, L., Respiratory cancer in Danish bakers: A 10 year cohort study. *Br. J. Ind. Med.*, 43, 516, 1986.

Vähäkangas, K. H., Samet, J. M., Metcalf, R. A., et al., Mutations of p53 and ras genes in radon-associated lung cancer from uranium miners. *Lancet*, 339, 576, 1992.

Wald, N. J., Nanchahal, K., Thompson, S. G., et al., Does breathing others people's tobacco smoke cause lung cancer? *Br. Med. J.*, 293, 1217, 1986.

World Health Organization histological typing of lung tumours. Second edition. *Am. J. Clin. Pathol.*, 77, 123, 1982.

Wu, A. A., Henderson, B. E., Pike, M. C., et al., Smoking and other risk factors for lung cancer in women. *JNCI*, 74, 747, 1985.

Wu-Williams, A. H., Dai, X. D., Blot, W., et al., Lung cancer among women in northeast China. *Br. J. Cancer*, 62, 982, 1990.

11

NONCARCINOGENIC RESPIRATORY DISEASE

Michael D. Lebowitz

CONTENTS

INTRODUCTION

This report evaluates non-carcinogenic respiratory diseases associated with environmental exposures, discusses impacts, and sets priorities for research in selected areas. The report attempts to focus primarily on respiratory problems in Europe. Recommendations will relate to further research, surveillance and assessment, and potential preventive programs.

The major environmental causes of respiratory diseases are air pollutants, primarily because of the inhalation route of entry. The respiratory outcomes to be considered include acute and long-term excesses of mortality, chronic respiratory diseases (CRD, including Chronic Obstructive Pulmonary Diseases [COPD], asthma, and lung function impairment, produced by long-term [total] exposures), acute exacerbations of such diseases, other acute respiratory illnesses (ARIs), and other responses. Acute and chronic effects have been associated with both of the major types of air pollution: the sulfur oxide (SO_x) and particulate (PM) complex arising from the combustion of sulfur-containing fossil fuels, and the photo-chemical oxidants formed in the atmosphere from the chemical reaction between precursor hydrocarbon compounds and nitrogen oxides which are largely related to motor vehicle emission. Certain metals have also been associated with ARIs and non-carcinogenic CRD: nickel, manganese, chromium, vanadium, and beryllium. Volatile organic compounds (VOCs) may also play a role in ARIs and non-cancer CRDs.

Evidence from different types of environmental epidemiological studies are considered; to a limited extent, their methodological advantages and disadvantages are discussed. As important are the quality of the exposure and risk assessments, and derived population impact estimates (NAS, 1985, 1991; WHO, 1983, 1986a). Discussion of co-variates and confounders will be included only as necessary (as this is not a primary topic). There will be discussion of interactions of the pollutants, which in collinear fashion or as complex mixes play special roles in respiratory responses, as to their interactions (and/or collinearity) with microorganisms (infectious or allergenic), with personal behavioral factors (e.g., smoking), and with meteorological phenomena.

The effects of air pollutants are related to the type of pollutant and its reactivity. The nature of the pollutant (size, chemical composition, hydroscopicity, volatility, etc.) is important in determining exposure-dose characteristics, including where in the respiratory tract the effect may occur. Meteorological factors are also important to consider in determining the environmental pathway of exposure as well as factors that might affect health directly. Information on the concentration of pollutants in the air needs to be coupled with time-activities data (Moschandreas, 1981; NAS, 1981, 1991; WHO, 1983), as the latter determine exposure and thus risk. Some pollutants affect the respiratory system through other routes of entry (e.g., some nitrogen oxides [NO_x], some VOCs such as carbon tetrachloride [CCl_4], paraquat and other pesticides, gamma radiation, aspirated material, neurotoxins, drugs, and microorganisms), and some pollutants are inhaled but have their major impact on other organ systems (e.g., carbon monoxide [CO], lead [Pb]) (WHO, 1989). These will not be considered in any detail.

Basic biological mechanisms involved relate to whether a pollutant is an irritant, an immunological sensitizer, a depressant of host defenses, or if it has specific pharmacokinetics and primary or secondary toxicity. Thus, host susceptibility and sensitivity are important also. Host susceptibility helps define sensitive subgroups and is an important factor in our considerations, though not discussed fully herein (Lebowitz, 1991; Lebowitz and Burrows, 1986). Sensitivity or susceptibility in some individuals can be to all irritants, allergens, or infectious agents, but such conditions are likely to be response specific or organ specific. Susceptibility typically implies that the individual is endowed with some biological or medical characteristic that may lead to an enhanced response (Lebowitz, 1991; Lebowitz and Burrows, 1986). The underlying characteristic may be shared by others in the population, though such subgroups are usually only a fraction of the general population. Such individuals require special evaluation and attention in all exposure-response studies and risk assessments (Lebowitz, 1991; Lebowitz and Burrows, 1986; Stebbings and Fogelman, 1979). Asthmatics are excellent examples of individuals who were susceptible to the disease, and once inflicted, are susceptible and sensitive to the effects of many environmental and non-environmental agents (Lebowitz and Spinaci, 1993; Infante-Rivard, 1993). Their susceptibility may have been innate (e.g., genetic) and/or induced by events/exposures in their life. Thus, the proportion of asthmatics in an exposed population is very critical.

This chapter will focus on specific pollutants for which sufficient information is available to assess potential population impacts. Primary reliance will be placed on the information contained in prior criteria documents, especially the WHO Air Quality Guidelines (AQG) for Europe (WHO, 1987). (This approach will be followed to the extent possible, given the following limits of the existing AQG: (1) they are for specific pollutants, and only one combination is considered; (2) they do not determine degree of severity of effects; and (3) they may, in places, require updating.) The seriousness/severity of the effects, acute and chronic, will help prioritize concerns. Seriousness will be determined by several factors, including disability, age of occurrence and subsequent sequelae, and relative economic impact. One must realize that inclusion of these disparate criteria can lead to some paradoxical, if not some contradictory, conclusions. The respiratory disease outcomes are evaluated in order of priority based on their severity/seriousness.

How serious are some of the respiratory effects? It is proposed that chronic respiratory diseases are the most severe of the health outcomes from air pollutants. This is especially the case if they occur in childhood, as one can argue strongly that such occurrences often lead to impaired lung function in life, to continuing symptomatology, and to adult chronic respiratory diseases

(Lebowitz and Burrows, 1986). Further, asthma is most likely the primary chronic respiratory disease of concern based on these criteria and the threat of early mortality (Lebowitz and Burrows, 1986; US NHLBI, 1991; US NIAID, 1979). Chronic symptomatology and low initial lung function also predict more rapid declines in function, and are associated with increased mortality; related mortality may be premature as well. Chronic respiratory diseases are associated with significant disability, often for decades; these start as early in life as the late forties. Such disability is associated with significant reductions in work days and work performance, as well as increased cost of medical/health care. The seriousness of the chronic respiratory diseases relates economically also to the acute exacerbations and subsequent disability experienced (US NCHS, 1988). There are also large psycho-social impacts of disability and of exacerbations at all ages (though especially in childhood). There is little doubt that a strong etiologic interaction of air pollution with smoking, work exposures (and other factors, such as alcohol consumption), and host factors, play a large role in the etiology and natural history of these diseases and their exacerbations (Lebowitz and Burrows, 1986; Newman-Taylor and Tee, 1990; Lebowitz and Spinaci, 1993; Lambert and Reid, 1970).

Low on the list of serious health outcomes are the transient functional and symptomatic impairments that occur in normal individuals when exposed to sufficient concentrations of irritants. It will become evident that we are unclear as to the role these have in the long-term vis-a-vis chronic respiratory diseases. Nevertheless, one can discuss their economic and psycho-social impacts.

MORTALITY

Normally, mortality is considered the most serious health effect following exposure to air pollution. This association is well known (ATS, 1978; WHO, 1987; US EPA, 1982, 1986; Lebowitz, 1983; Waller and Swan, 1993): acute mortality episodes have been associated with large increases of PM and SO_2 (see Tables 1–3). Excess mortality usually occurs in these acute air pollution episodes, with poor weather (such as occurred in the 1948 Donora episode, and to a great extent in the London episodes of 1952 and later) (ibid.). Increased mortality due to long-term exposures to similar reducing-type atmospheres (primarily SPM and SO_x at lower concentrations for longer periods of time) also occurs (ibid.).

Variations in PM below the guideline levels (500 $\mu g/m^3$) have been associated with 5% of the variation of daily mortality in New York City (US EPA, 1982; Buechley et al., 1973; Lebowitz et al., 1973a); and associations of daily variations in sulfur oxide/particulate air pollution of lesser magnitude have also been studied with respect to fluctuations in daily mortality in other

Table 1   Health Effects and Dose/Response Relationships for Particulates and Sulfur Dioxide

Average Time for Pollution Measurements	Location	Particles mg/m³	SO₂ mg/m³	Effect
24 hr	London	2.00	1.04	Mortality
		0.75	0.71	Mortality
		0.50	0.40	Exacerbation of bronchitis
	New York City	5 COHS [a]	0.50	Mortality
		3 COHS	0.70	Morbidity
	New York City	0.145 (+?)	0.286	Increased prevalence of respiratory symptoms
	Birmingham	0.18–0.22	0.026	Increased prevalence of respiratory symptoms
	New York City	2.5 COHS	0.52	Mortality
	Steubenville	0.036–0.209	0.007–.055	Mortality
Weekly mean	London	0.20	0.40	Increased prevalence or incidence of respiratory illnesses
6 Winter months	Britain	>0.10	>0.10	Bronchitis sickness absence from work
Annual	Britain	0.07	0.09	Lower respiratory infection in children
		0.10	0.12	Respiratory symptoms and lung function in children
	Buffalo	0.08	0.045	Mortality
	Berlin	0.18	0.73	Decreased lung function

[a] Coefficient of haze units.

Data from ATS, 1978; EPA 1982a–b; Lebowitz, 1983, WHO, 1987; Phalen, 1995.

metropolitan areas, including London, Oslo, Philadelphia, Los Angeles, Tokyo, Osaka (WHO, 1987; US EPA, 1982; Lebowitz et al., 1973b; Lebowitz, 1983; Pope et al., 1992; Schwartz et al., 1992; Waller and Swan, 1993). Analyses of several sets of data tend to agree in finding greater mortality (all causes and cardio-pulmonary), especially in those over age 45, on days of greater pollution. Fine particulates and sulfate are the pollutants most closely associated with cross-sectional mortality increases. About 4% of geographic differences in cardio-pulmonary mortality is associated with PM and sulfate (with socio-economic status [SES] controlled) (US EPA, 1982; WHO, 1987).

"On the basis of current exposures, and observed mortality during earlier London episodes, present best estimates point to the likelihood of several thousand excess deaths per year due to winter [reducing]-type episodes" (WHO, 1991). Many intervening factors, such as temperature extremes,

Table 2   Exposure-Effect Relationships of Sulfur Dioxide, Smoke, and Total Suspended Particulates: Effects of Long-Term Exposures ($\mu g/m^3$)

Sulfur Dioxide	Smoke	Total Suspended Particulate	Location	Effects
200	200	—	Sheffield, England	Increased respiratory illness in children
—	—	180[a]	Berlin; NH, U.S.	Increased respiratory symptoms, decreased respiratory function in adults
150	—	—	England and Wales	Increased respiratory symptoms in children
140[b]	140[b]	—	Great Britain	Increased lower respiratory tract illnesses in children
60–140[c]	—	100–200[d]	Tokyo, Japan	Increased respiratory symptoms in adults
50	50	—	WHO	Chronic lung disease

Note: Concentrations (annual means of 24-hr mean values, $\mu g/m^3$).

[a] High volume sampler (2-month mean, possible underestimation of annual mean).

[b] Estimates based on observations after end of study probable underestimations of exposure in early years of study.

[c] Automatic conductimetric method.

[d] Light scattering method, results not directly comparable with others.

Data from ATS, 1978; EPA, 1982b; Lebowitz, 1983; WHO, 1987.

influenza epidemics, holiday weekends, and season of the year, have strong effects on the day-to-day number of deaths and may enhance or minimize the effect of air pollution. However, there is no agreement as to exactly how many deaths may be attributed to air pollution. Tables 1–3 illustrate the results of the more important studies on the association of mortality with air pollution.

The psycho-social impact of the increase mortality, especially that due to premature mortality, is always great. If some excess mortality or morbidity does occur in children, then this type of effect requires further consideration (Phalen, 1995).

Recent studies, primarily in the U.S., have appeared to show effects of lower levels of PM on mortality (Pope et al., 1992; Schwartz et al., 1992; Phalen, 1995), but they have not been replicated. Reviews and editorials have focused on these issues, but have not concluded that such evidence is biologically plausible or convincing (Waller and Swan, 1993; Utell and Samet, 1993; Lebowitz, 1995).

Table 3 **Exposure-Effect Relationships of Sulfur Dioxide, Smoke, and Total Suspended Particulates: Effects of Short-term Exposures**

Sulfur Dioxide	Smoke	Total Suspended Particulate	Location	Effects
>1000	>1000	—	London, 1952	Very large increase in mortality to about 3 times normal, during 5-day fog, pollution figures represent means for whole area: maximum (central sites) sulfur dioxide 3700 $\mu g/m^3$
710	750	—	London, 1958–1959	Increases in daily mortality up to about 1.25 times expected value
500	500	—	London, 1958–1960	Increases in daily mortality (as above) and increases in hospital admissions becoming evident when pollution levels shown were exceeded (magnitude increasing steadily with pollution)
500	—	—	New York, 1962–1966	Mortality correlated with pollution: 2% excess at level shown
500	250	—	London, 1954–1968	Increase in illness score by diary technique among bronchitic patients seen above pollution levels shown (means for whole area)
200[a]	—	150[b]	Cumberland, WV	Increased asthma attack rate among small group of patients, when pollution levels shown were exceeded
—	—	80	Tucson, AZ	

Note: Concentrations (24-hr mean values, $\mu g/m^3$).

[a] West-Gaeke method.
[b] High-volume sampling method.

Data from ATS, 1978; EPA, 1982b; Lebowitz, 1983; Lebowitz et al., 1985.

CHRONIC RESPIRATORY DISEASES

There is no doubt that the long-term AQG (annual average) for suspended particulate matter (SPM), which does not specifically differentiate subfractions (i.e., total suspended particulate [TSP] vs. particulate matter <10 μm aerodynamic diameter [PM10]) or chemical composition, is based on the association between TSP and chronic obstructive pulmonary diseases (COPD), primarily chronic bronchitis and impaired lung function; the association is well documented (WHO, 1987; US EPA, 1982, 1986b; ATS, 1978; Lebowitz, 1983, 1995; Abbey et al., 1993, 1995). Although PM10 and suspended sulfate (SS,

SO_4) may be more important than TSP because of deposition characteristics, TSP also is associated with COPD; discussions as to why this happens (primarily particle size characteristics) have been dealt with elsewhere (Lebowitz, 1988b; Leaderer et al., 1977, 1986; Koenig et al., 1990b). An important exposure to SPM may occur indoors as well as outdoors and may be related to environmental tobacco smoke (ETS), other combustion sources, and poor ventilation (NAS, 1981, 1991). To some extent sulfates (SO_4) and nitrates also play a role indoors (US EPA, 1974, 1982; Lebowitz, 1988).

What is left somewhat unsettled is the contribution of long-term exposures to sulfur dioxide (SO_2) and acid aerosols: these pollutants may occur in a collinear fashion with others in specific locations, dependent on source and long-range transport.

The WHO EURO AQG (WHO, 1987) also considers sulfuric acid as a probable causal agent of respiratory disease. Sulfur oxide (SO_x) species may be highly correlated, even without much SPM, around specific point sources; they usually are collinear with SPM in urban areas, though SPM can be high without high SO_x.

There is little doubt that tobacco smoking, and some workplace exposures to SPM, which are more important in disease etiology, also interact with outdoor SPM in producing COPD; the recent EPA report on ETS discusses these issues (US EPA, 1992). Other factors of likely importance, which are time-place dependent, include family history, alcohol consumption, SES, age and duration of exposure (with greater effects in children), and other urban factors (Lambert and Reid, 1970). The study of Abbey et al. is one of the most recent efforts demonstrating such chronic effects in a multivariate fashion (Abbey et al., 1993, 1995; Robbins et al., 1993).

It is unclear whether ozone (O_3) (or other oxidants) produce COPD with long-term high exposures; animal models have shown either obstructive changes or restrictive (fibrotic-type) changes, and reversibility is not well understood yet. At least one study has shown an association of diagnosed chronic bronchitis (as well as asthma and bronchial responsiveness, especially in low SES subgroups) in childhood with indoor exposure to aldehydes, ETS (especially in the interaction of the two), and NO_2 (Krzyzanowski et al., 1990; Lebowitz et al., 1989; Quackenboss et al., 1989). A few other studies appear to find a relation between either ambient SPM or ETS with asthma as well (relative risk [RR] of 1.5–2.5) (US EPA, 1992). It is also possible that the hydrocarbons found in diesel smoke (as has been shown in mining), in unvented kerosene heaters, in auto exhaust, or in other possible sources, play an important role (Von Mutius et al., 1992; NAS, 1993; Blanc et al., 1993; Leaderer et al., 1986; Corbo et al., 1993; Zwick et al., 1991).

Nitrogen dioxide (NO_2) can produce a reversible form of COPD in animals (ATS, 1978), but is probably not important as a cause of COPD at concentrations to which humans are exposed. However, it was shown in some studies that NO_2 at ambient concentrations increases bronchial responsiveness, a key feature of asthma, especially in those subjects with low SES (Bylin et al., 1988; Quackenboss et al., 1989). Cadmium is also associated with COPD; and other metals appear to be sensitizers, producing bronchial responsiveness (Ni, Cr, V, platinum salts) (ATSDR, 1991; WHO, 1983; 1987). Biological and chemical aero-allergens can cause asthma but are unrelated to COPD (Sporik et al., 1990; NAS, 1993). Pulmonary fibrosis can be produced by some of these metals and by asbestos (WHO 1983; ATSDR, 1991; ATS 1978). Other air pollutants do not appear to be important factors in producing non-carcinogenic CRD.

Decrements in pulmonary function in children related to air pollutants are well known (WHO, 1987; ATS, 1978; Lebowitz, 1983). These decrements may be precursors to COPD.

Exposure-Response Relationships

COPD symptoms/diagnoses appear to increase significantly (RR of 1.5–2.5) as annual TSP and/or SO_2 increase above 100 $\mu g/m^3$ (Table 1). The WHO (1987) Air Quality Guidelines quote a study by Kitegawa where SO_2 concentrations of about 159 $\mu g/m^3$ (based on SO_3 of about 130), with peaks up to 100 times greater, were responsible for 600 cases of allergic bronchitis; pollution control measures reduced the number of new cases. Leaderer et al. (1977) showed that chronic bronchitis appears to increase linearly with SO_4 increase, after adjustment for gender and smoking habits: every 2 $\mu g/m^3$ increase adds 1.24% to the prevalence rate for both sexes; using step functions in data analysis, he suggested a level of 5.8 $\mu g/m^3$ sulfates as the point at which chronic bronchitis starts increasing rapidly. Abbey et al. (1993) found that an average annual increase in SO_4 of 7 $\mu g/m^3$, controlling for other pollutants, yielded a relative risk of 2.85 for asthma incidence; the relative risk for chronic bronchitis incidence associated with SO_4 was lower (1.43) and not significant, but significant increases were found related to TSP and PM10 (Abbey et al., 1995). In urban areas, significantly more chronic COPD symptoms may occur with SO_4 above 9 $\mu g/m^3$ in the presence of high SO_2 and TSP (WHO, 1987). A 24-hr AQG of 180 $\mu g/m^3$ of NO_2 was established to avoid potential chronic effects from repeated exposures (WHO, 1987).

Pulmonary Function Declines

Significant decrements (3–8%) of pulmonary function are associated to ambient annual TSP above 180 $\mu g/m^3$ or PM10 of about 110 or 100 $\mu g/m^3$ of SS and SO_2 in children (Table 1). Significant differences (< 3%) occur in children in relationship to ETS exposure (mostly due to PM2.5) of 60–100 $\mu g/m^3$ or more (US EPA, 1992). Decreases in lung growth (FEV [Forced Expiratory Volume in 1 s] differences of <3%) are seen with PM and with ETS, but not with SO_2 (US EPA, 1982, 1992; WHO, 1987; Lebowitz et al.

1992a). Decreases occur more and are larger in those starting with low lung function, bronchial responsiveness, and/or a chronic respiratory disease (Sherrill et al., 1992; Taylor et al., 1985).

EXACERBATIONS OF CHRONIC RESPIRATORY DISEASE

Almost all the major pollutants can produce exacerbations of CRD (COPD and asthma), and temporal declines in pulmonary function. The exacerbations produced by the reducing-type pollutants (PM and SO_x) are best documented (Tables 1–3). An extensive series of studies on the effects of air pollution on bronchitic patients were conducted in Britain between 1955 and 1970 (Lawther et al., 1970, 1974a-c). They show that exacerbations of the disease were associated with high concentrations of smoke and SO_2. Aggravation of the disease was more associated with relative increases rather than absolute concentrations. Further, in the U.K. review of sickness absence records, of rates of physician consultation, and of daily records of hospital admissions through the emergency service showed associations with periods of heavy air pollution (Lawther, 1963). With decreasing concentrations of pollutants in Britain, it has been difficult (since 1969) to relate subjects' symptom status to variations in air pollution. Epidemiological research indicates that the transformation products of sulfur dioxide in ambient air, principally sulfates, are more likely than sulfur dioxide alone to be responsible for many of the adverse health effects typically associated with sulfur oxides. Studies conducted in several North American areas suggest that high daily or annual sulfate levels are associated with increased attack frequencies in asthmatics, some symptoms in cardio-pulmonary patients, decreased ventilatory function in schoolchildren, and symptoms of acute and chronic respiratory diseases in children and adults (Finklea, 1974; Bates and Sizto, 1983).

Asthma actually has many triggers, most of which are environmental, including PM/ETS, SO_2, ozone, NO_2, formaldehyde (HCHO, probably with particles), and aeroallergens (WHO, 1986, 1987; NAS, 1981, 1993; ATS, 1978; Albertini et al., 1989; Hanley et al., 1992; Horstman et al., 1986; O'Rourke and Lebowitz, 1995; Cohen et al., 1972; Forastiere et al., 1994; Kagamimori et al., 1986; Blanc et al., 1993; Seltzer et al., 1986; Molfino et al., 1991; Whittmore and Korn, 1980; Zagraniski et al., 1979). Airborne infectious agents, especially those producing lower respiratory illnesses (LRIs), are also a major cause of exacerbations in asthma (and bronchitis), and illness in children (Lancet, 1992; Horn and Gregg, 1973). Hospitalizations for asthma and CRD are definitely associated with air pollution, especially PM, SO_x (as shown in recent studies in Spain, Finland, and the Netherlands), and O_3 (Bates and Sizto, 1983; Schwartz et al., 1993; Roemer et al., 1993; Sunyer et al., 1993; Ponka et al., 1991). One should assume that exceedances of the AQGs for any short-term irritant would produce such exacerbations (WHO, 1987).

As shown in most studies, low temperatures can often exert a greater effect than air pollution, and even high temperatures can affect asthmatics (ATS, 1978; US EPA, 1982, 1984, 1986; Lebowitz et al., 1992a,b; Rossi et al., 1993; Cohen et al., 1972).

Exposure-Response Relationships

The correlation of chronic bronchitis exacerbations with SO_2 levels, when its mean is above the AQG in winter, is low (r value around 0.3) (WHO, 1987, 1991; Lawther et al., 1970). One can assume from the available data that increases of PM and SO_2 to concentrations above 250 $\mu g/m^3$ augment the rate of exacerbations by 1.3–1.8 times, and concentrations above 500 $\mu g/m^3$ increase the rate by 1.8–2.6 times; higher rates occur in the lower SES groups (WHO, 1987, 1991; Lebowitz, unpublished).

Attributable risk estimates associated to exposure to respirable PM are in the range 37–63%. Other findings include a significant decrease in lung function in exercising asthmatics at 100 $\mu g/m^3$ of sulfuric acid or 715–2145 $\mu g/m^3$ SO_2 (WHO, 1987). Ozone, alone and in combination with TSP and/or temperature (relatively low or high), has been shown to exacerbate asthma and chronic bronchitis at levels above 160–200 $\mu g/m^3$ (WHO, 1987; Lebowitz et al., 1985; Lebowitz, unpublished).

OTHER ACUTE CONDITIONS

Acute respiratory illnesses (ARI) are increased by 1.5–2.0 times with exposures to PM (including ETS), SO_2, NO_2 (US EPA, 1982, 1986, 1992; WHO, 1987, 1991; Lebowitz, 1988a-b; Moseholm et al., 1993). Early childhood ARIs also increased by 1.5 times; the increase is 2.5 times in subjects of lower SES (Lebowitz, 1988b) and appears to be associated to SS and SO_2 concentrations of 190 $\mu g/m^3$. Hospitalizations due to ARIs are also increased by 1.5–2.8 times when air pollution episodes occur (Douglas and Waller, 1966; Lunn et al., 1970).

The level of NO_2 reputed to produce acute respiratory illnesses is about 137 $\mu g/m^3$ (1-hr levels) (WHO, 1987). In an international study, the weekly mean of NO_2 concentration in homes with gas cookers was 105 $\mu g/m^3$, though values higher than 950 $\mu g/m^3$ were found with intense cooking (1–6 hr) (WHO, 1987; RIVM, 1984; CEC-COST, 1989); similar results were found recently in the U.S. six-city study (Neas et al., 1991). One study reported an increase in acute respiratory diseases in families with 24-hr exposure to ambient mean NO_2 of 150–282 $\mu g/m^3$ (suspended nitrate level of 3.8 $\mu g/m^3$) and no other associated pollutants (Braun-Fahrlander et al., 1992).

Ozone/oxidants may increase ARIs also, but no level of effect has been determined. Acute respiratory symptoms are almost always associated with increased ozone exposures (>200 $\mu g/m^3$) in field studies (WHO, 1987).

Pulmonary function decrements occur in normal children and adults with exposure to 110 μg/m³ of PM10 (in the presence of SO_2), 3760 μg/m³ of NO_2, above 150–200 μg/m³ of ozone for 1 hr (above 100–250 for 8 hr) (WHO, 1987; Pope et al., 1991). Exercise enhances dose and thus pulmonary responses. Acute symptoms (e.g., cough, phlegm, chest tightness, dry or sore mucous membranes) are almost always present at levels producing significant changes in function.

Acute pulmonary function decrements have been found in children and asthmatics in field studies at levels lower than those cited, assumed to be due to more continuous (i.e., cumulative) exposure and to cumulative effects of other pollutants, as well as to the use of more sensitive measures of change (e.g., PEF [Peak Expiratory Flow]) (Krzyzanowski et al., 1992; Lebowitz et al., 1985; Lioy et al., 1985; Lippmann, 1989; Rombout et al., 1986; Blanc et al., 1993). Also, field studies have found more consistent and significant bronchial responsiveness associated with NO_2 than might occur in chamber studies (Quackenboss et al., 1991). Thus, in the real-life situation, effects will probably occur at levels below those indicated by the AQG, and will certainly occur in polluted areas.

The combined effect of all contaminants, especially the combination of infectious agents and indoor air pollution, is thought to account for a substantial proportion of absenteeism in schools and workplaces (Lebowitz, 1983; Romieu et al., 1992), and of days of restricted activity or performance; 3–5 days of such restricted activity per person per year may result specifically from these environmental effects (WHO, 1990).

ESTIMATED POPULATION IMPACTS

Chronic Respiratory Disease

Western Europe has annual levels of SO_2 and SPM below the AQG level of 100 μg/m³, though some urban areas exceed the SPM AQG (e.g., cities in Southern Europe: Milan [161–180 μg/m³], Lisbon [110], Athens [SS >120, up to 600]) (WHO, 1991). Large areas of eastern Europe exceed the level of 100 μg/m³ (by 2–4 times): in former Czechoslovakia (up to 140 μg/m³), 50% of former East Germany, Southern Poland (100–300 μg/m³), and also urban areas in Eastern Europe (e.g., Zagreb [150], Bucharest [110], Cracow [170]) (WHO, 1991, 1994; GEMS, 1982). About 152 million people live in urban areas of Southern and Eastern Europe (WHO, 1991). If prevalence rates of COPD in Western Europe are about 6%, then with a RR of 1.5–2.5 associated to air pollution, an excess of 3–9%, or 4.5–13.5 million, cases can be expected; even half of this excess (2–7 million additional cases) would be considered appreciable.

In Eastern Europe, for example, in Cracow, a city with high air pollution level, the prevalence rate of "chronic bronchitis" was 24% in men (64% in current smokers, 14% in ex-smokers) and 11.5% in women (22.5% in current smokers, 2% in ex-smokers) at end of a 13-year follow-up (Krzyzanowski et al., 1990). The change in the prevalence rate of asthma is equivalent. In Cracow, the asthma syndrome prevalence rate was about 8%, while it is about 5–6% in the U.S. and Western Europe. The economic impact of such excesses must be in the billions of U.S. dollars (US NCHS, 1988; US NIAID, 1979; US NHLBI, 1991; NAS, 1993).

Exacerbations of Chronic Respiratory Disease

A US NIAID (US NAID, 1979) report stated that asthma is among the leading causes of physician visits, hospitalizations, and workdays lost. The exacerbation rate in asthmatics is well known and it is based on data from emergency room and hospital visits. If in the heavily polluted areas the incidence rate of the disease is increased by 1.5 and the exacerbations are doubled in those with the disease, the overall rate of disease episodes is increased by threefold. In the U.S., the annual cost for asthma was 6.2 billion U.S. dollars in 1990 (Weiss et al., 1992).

One should assume that the prevalence rate of all airway obstructive diseases (COPD, including impaired pulmonary function, and asthma) is about 18–24% (US NCHS, 1988; US NHLBI, 1991). Rates are much higher in those who smoke and are exposed to occupational hazards, especially in those with low SES groups (up to 80% in male workers who smoke). Even assuming an overall rate of 20%, about 30 million cases and 60–100 million exacerbations per year could be expected in the 152 million people living in heavily polluted urban areas in Europe (WHO, 1991).

The number of exacerbations has also been influenced also by the photo-oxidant pollution, mostly in summertime. If the 1-hr ozone concentration of 160–200 $\mu g/m^3$ or above and 8-hr values of 100–120 $\mu g/m^3$ can produce exacerbations, then Western European (and probably also Eastern European) asthmatics and bronchitics have reason for concern during exceedances (1–54 days above 1-hr AQG in low O_3 years, 10–90 days above 1-hr AQG in high years, 10–100 days above 8-hr AQG) (WHO, 1991). Southern European metropolitan areas have many more exceedances; for instance, Athens' air is above 300 $\mu g/m^3$ on many days (WHO, 1991); highly industrialized areas of northern Italy often exceed the AQG, with the highest single 1-hr value reported (1990) as 550. Several rural areas in Germany also have half-hour values above 180. Thus, exacerbations will occur throughout Europe, usually on 10–100 days, mostly in the summer. We have to assume that the total number of exacerbations is increased with the addition of summer exceedances in Western and Northern Europe. If there were 250–265 million living in those metropolitan areas and the prevalence rates were 6% for asthma, then 15–16 million asthmatics would be assumed to have at least one exacerbation in summer. Thus, one would have to add 15 or more million exacerbations to the number previously estimated, for a total between 75–120 million exacerbations.

Other Acute Illnesses

Absenteeism from work and school is associated to ARIs. For instance, out of the 152 million people living in highly polluted areas in Europe, it could be estimated that 137 million people (those with age 5+) would have at least 3 days of absenteeism per year, i.e., 411 million person-days.

A WHO report (WHO, 1994) indicates 10% of the Western European population is exposed to indoor concentrations of NO_2 higher than the 1-hr AQG. Thus, about 34 million will probably experience a 2–4 excess of ARI episodes a year.

If 80–90% of European homes has at least one smoker, and if one assumes 50% as lower (vs. higher) SES, then about 30–45% of all ARIs probably has a significant ETS contribution.

As stated in a WHO document, "The 1-h peak O_3 concentrations in these areas [northwestern, eastern, urban areas in southern Europe] range between 200–500 µg/m^3. An 8-h AQG value will presumably be exceeded by 2–4 times. [summer-type episodes]" (WHO, 1991). Thus, acute symptoms and pulmonary function decrements will occur throughout Europe on at least 10–100 days in the summer.

RECOMMENDATIONS

Short-Term Plans (1 year)

The first action should be to extend the data base to have monitoring, source apportionment, and dispersion modeling, results for various areas of Europe for whatever pollutants there are data in order to obtain regional exposure estimates. These data should be coupled to population data for the same areas to form a matrix to estimate how many are exposed to how much pollution.

The second action should be to use the data with known respiratory effects (exposure-response) information to determine preliminary population impacts. To further this effort, prevalence rates of chronic respiratory diseases, LRIs, ARIs, and so forth should also be obtained to improve the risk assessments for sub-populations of susceptible and sensitive individuals.

The third action would be to estimate economic impacts and determine short-term preventive measures based on cost-benefit ratios.

Long-Term Plans

Pollutant Assessment

Outdoors, PMs generated by various sources are complex pollutants to measure and require monitoring, chemical analysis, and modeling (including source apportionment). Indoors, PMs generated by cigarette smoke, wood stoves and fireplaces, and other combustion are complex to study and contribute a major portion of the total PM exposure. Further, methods for evaluating adsorption or absorption of gases onto particulates are necessary, since these different combinations may have very different effects, some of which are more serious than the gas or PM alone. The size distribution of PM with different combinations of sources requires further evaluation, as the PM will produce major responses dependent on both its size and chemical composition. Thus, characterization of the sizes and types of PM occurring both indoors and outdoors is needed to evaluate the independent effects on health of PM and the interaction of PM with other pollutants in producing effects on health.

Exposure Assessment

Monitoring and time-activity diaries should directly assess levels of exposure (over time) and help to identify the major determinants of personal exposure of individuals to the pollutants. Quantitative assessments of total personal exposure are possible even with fixed location (microenvironmental) sampling coupled with individual time budgets. Exposure classifications or models alone are probably insufficient to characterize exposure patterns of individuals. Some direct monitoring component (indoor, outdoor, and personal) is required to determine the magnitude and causes of misclassification errors.

Some exposure assessment methods have already been established (e.g., for TSP, SO_2, NO_2, O_3). Most assessments for acid precipitation, fine particulates, sulfates, and nitrates will be unsuccessful because most of the information required is unavailable. Particle deposition per se is fairly well understood. Sizing and speciation of wet and dry particulates of interest need further work.

Although peak concentrations during source use are not directly available from integrated samplers, estimates of the average concentration during source generation can be derived from the reported or measured total time of source use and the measured integrated average; this approach requires further verification with continuous monitoring data.

Exposure-Response Studies

There are still serious limitations in our knowledge of total population exposure for each contaminant, and equally serious limitations in our knowledge of the exposure-response relationships, either to a single or to multiple contaminants. Insufficient monitoring and associated studies of outcome are lacking in major parts of Europe.

The expected time interval between exposure(s) and the eventual health effect must be taken into account. Further, patterns of response, and the interaction between pollutants and other factors (such as socioeconomic characteristics, smoking, and occupational exposure) can only be shown in specific population studies.

Risk Assessment

For risk assessment, one requires an exposure-effect relationship, an estimation of the size of the population at risk (exposed), including the groups especially sensitive, good exposure information, appropriate data on health outcomes, and some reasonable future scenarios. Better models of exposure and exposure-response must be developed using real data to meet the needs of criteria for air quality standards.

Hazard assessment is made for populations. It is complicated by the varied biological processes that may result from exposures, usually multiple. The health hazard assessments differ sufficiently with the different mechanisms (allergic, infective, or irritant/toxic). Each model has parameters determined by the exposure, time, and proportion of susceptibles. These distributions of compounds can represent the risk assessment exposure-response relationships; they can be defined for different populations.

Biological aerosols are different from other types of airborne particulate matter, as they have complex, varied organic structures that do not permit simple sampling or chemical analysis. Even in small quantities, they can have very large respiratory impacts due to concomitant infection, allergenicity, or irritation. These health effects can range from uncomfortable to disabling. Only in sensitive individuals would other air contaminants have such allergic or irritant effects. Temperature and humidity contribute to growth of some organisms that produce such particles, as well as affecting health directly. Bacteria and fungi can multiply in microclimates indoors, including ventilation systems. Viruses do not multiply indoors but may spread between humans and from a few animal sources. The airborne spread of viral disease can be facilitated through crowding of people, or through the spread of airborne virus through the ventilation system. Some bioaerosols infiltrate from outdoors (e.g., pollen). Facilitation of pathogenicity also occurs with prior (and continued) exposure to other air contaminants (e.g., volatile organic compounds [VOCs] and NO_2). Bioaerosols have a large attributable risk for respiratory effects. Thus, risk assessment of these contaminants is also necessary.

Abatement and Prevention

Strategies for prevention against the effect of airborne contaminants are of two major types, those relating to individuals and those relating to the physical environment (WHO, 1986b; Samet, 1990; Lebowitz, 1993). Prevention for the individual can occur through modification of host susceptibility (e.g., immunization for infectious agents) and avoidance of irritants or allergen pollutants. Avoidance of triggering exposures (as recommended clinically to asthmatics) including physical measures is predominantly based on the avoidance of conditions, and on the containment and removal of sources. Society can exert considerable pressure, or provide/promote incentives to control sources of contamination. Education could include the dissemination of information to individuals on actions they could take to reduce/eliminate their own sources of pollution, and the ability to recognize situations (e.g., source use) which might contribute to heavy concentrations. It is recognized that use of the information will depend upon the available resources.

Physical Methods

The goal of protecting human health is not only served by the application of outdoor air quality standards. The experience gained in developing and implementing strategies for population exposure reductions in the outdoor environment is very useful. New strategies will need to be considered, developed, tested, and evaluated for indoor environments. These methods of source control have been categorized by a WHO working group (WHO, 1990). Proper running of industrial processes and design and construction of buildings are the most desirable control means to avoid contamination from their system-related sources. Strategies should avoid the conflicts between energy conservation and effective control. Source modification offers several possibilities for control.

Economics

The financial burden associated with indoor as well as outdoor controls should be adequately estimated when promoting preventive measures.

Guidelines and Standards

To protect the public, some governmental guidelines for reduction of contamination have been established in some countries but need further development in other countries. A strategy should be pursued as part of public policy, which should include where necessary development of standards, regulations, controls, practices, labeling requirements, recommendations, and guidelines (Lowrance, 1976; WHO, 1986, 1987). Public health-based standards should provide adequate protection for susceptible groups (ATS, 1978; WHO, 1987). Results have been positive, in general, when pollutant concentrations were relatively high, as in London and New York (see Tables 1–3). Governmental support of research and communication can be a primary service to encourage cooperation among participants in prevention.

RESEARCH OBJECTIVES

Chronic Effects of Multipollutant Exposures

This objective includes: (1) differentiation of susceptibility to chronic effects, through the ongoing studies on bronchial responsiveness, immunological and physiological status, and co-existing morbidity; (2) determination of quantitative and qualitative response differences, and differences in response time, for susceptible individuals (and comparisons); (3) the study of the roles of the natural history of responses to climate, biological and nonbiological aerosols in different subpopulations as factors in subchronic and chronic responses; and (4) the study of specific pollutant interactions in chronic exposure-response relationships.

Acute Effects of Multipollutant Exposures

The objectives are to relate the respiratory responses (symptoms, peak expiratory flows, spirometry, medication usage, and medical visits) and changes in activity/performance to primary air pollutants to aeroallergen exposures, and to meteorological conditions, as modified by host characteristics (age, sex, SES, reactivity or predisposition to response) and by time/activity patterns in both indoor and outdoor environments. Elevated short-term concentrations of pollutants (that are also related to intermittent source use indoors, such as smoking or combustion appliances) should also be of interest in assessing the health effects of these pollutants. (For example, ETS and NO_2 may facilitate infectious processes and bronchial responsiveness, especially in children; and these infections may lead to more serious sequelae.)

REFERENCES

American Thoracic Society (ATS), *Health Effects of Air Pollution*, New York, 1978, 1989.

Anonymous, Indoor air pollution and acute respiratory infections in children (Editorial), *Lancet*, 339:396, 1992.

ATSDR (USA Agency for Toxic Substances and Disease Registry), *Toxicological Profiles*, USATSDR, Atlanta, 1991.

Abbey, D. E., Lebowitz, M. D., Mills, P. K., Petersen, F. F., Beeson, W. L., and Burchette, J., Long-term ambient concentrations of particulates and oxidants and the development of chronic disease in a cohort of non-smoking California residents, *Inhal. Toxicol.*, 7:19, 1995.

Abbey, D. E., Petersen, F. F., Mills, P. K., and Kittle, L., Chronic respiratory disease associated with long-term ambient concentrations of sulfates and other pollutants, *J. Expos. Anal. Environ. Epidemiol.*, 3 (S1):99, 1993.

Albertini, M., Politano, S., Berard, E., Boutte, P., and Mariani, R., Variation in peak expiratory flow of normal and asymptomatic asthmatic children, *Ped. Pulmonol.*, 7:140, 1989.

Bates, D. V. and Sizto, R., Relationship between air pollution levels and hospital admissions in Southern Ontario, *Can. J. Public Health*, 74:117, 1983.

Blanc, P. D., Galbo, M., Hiatt, P., Olson, K. R., and Balmes, J. R., Symptoms, lung function, and airway responsiveness following irritant inhalation, *Chest*, 103:1699, 1993.

Braun-Fahrlander, C., Ackermann-Liebrich, U., Schwartz, J., Gnehm, H. P., and Rutishauser, M., Air pollution and respiratory symptoms in preschool children, *Am. Rev. Resp. Dis.*, 145:42, 1992.

Buechley, R. W., Riggan, W. B., Hasselbald, W., and Van Bruggen, J. B., SO_2 levels and perturbations in mortality: a study in New York-New Jersey metropolis, *Arch. Environ. Health*, 27:134, 1973.

Bylin, G., Hedenstierna, G., Lindvall, T., and Sundin, B., Ambient nitrogen dioxide concentrations increase bronchial responsiveness in subjects with mild asthma, *Eur. Respir. J.*, 1:606, 1988.

CEC-COST, Indoor Pollution by NO2 in European Countries, (Rep. #3), 1989.

Cohen, A. A., Bromberg, S., Buechley, R. W., Heiderscheit, L. T., and Shy, C. M., Asthma and air pollution from a coal fueled power plant, *Am. J. Public Health*, 62:1181, 1972.

Corbo, G. M., Forastiere, F., Dell'Orco, V., Pistelli, R., Agabiti, N., DeStefanis, B., Ciappi, G., and Perucci, C. A., Effects of environment on atopic status and respiratory disorders in children, *J. Allergy Clin. Immunol.*, 92:616, 1993.

Douglas, J. W. B. and Waller, R. W., Air pollution and respiratory infection in children, *Br. J. Prev. Soc. Med.*, 20:1, 1966.

Finklea, J., Health Consequences of Sulfur Oxides, EPA 650/1-74-004, 1974.

Forastiere, F., Corbo, G. M., Pistelli, R., Michelozzi, P., Agabiti, N., Brancato, G., Ciappi, G., and Perucci, C. A., Bronchial responsiveness in children living in areas with different air pollution levels, *Arch. Environ. Health*, 49:111, 1994.

GEMS, Global Environmental Monitoring System, Estimating human exposure to air pollutants, World Health Organization, Geneva, 1982.

Hanley, Q. S., Koenig, J. Q., Larson, T. V., Anderson, T. L., Van Belle, G., Rebolledo, V., Covert, D. S., and Pierson, W. E., Response of young asthmatic patients to inhaled sulfuric acid. *Am. Rev. Respir. Dis.*, 145:326, 1992.

Horn, M. and Gregg, I., Role of viral infection and host factors in asthma and chronic bronchitis, *Chest*, Suppl. 63:44S, 1973.

Horstman, D., Roger, L. J., Kehrl, H., and Hazucha, M., Airway sensitivity of asthmatics to sulfur dioxide, *Toxicol. Ind. Health*, 2:289, 1986.

Infante-Rivard, C., Childhood asthma and indoor environmental risk factors, *Am. J. Epidemiol.*, 137:834, 1993.

International Programme on Chemical Safety (IPCS), *Guidelines on Studies in Environmental Epidemiology*, (Environmental Health Criteria n. 27), WHO, Geneva, 1983.

Kagamimori, S., Katoh, T., Naruse, Y., Watanabe, M., Kasuya, M., Shinkai, J., and Kawano, S., The changing of respiratory symptoms in atopic children in response to air pollution, *Clin. Allergy*, 16:299, 1986.

Koenig, J. Q., Prior exposure to ozone potentiates subsequent response to sulphur dioxide in adolescent asthmatic subjects, *Am. Rev. Respir. Dis.*, 141:377, 1990a.

Koenig, J. Q., Hanley, Q. S., Rebolledo, V., Dumler, K., Larson, T. V., Wang, S., Checkoway, H., Van Belle, G., and Pierson, W. E., Lung function changes in young children associated with particulate matter from wood smoke, *Am. Rev. Respir. Dis.*, 141:A425, 1990b.

Krzyzanowski, M., Camilli, A. E., and Lebowitz, M. D., Relationships between pulmonary function and changes in chronic respiratory symptoms — comparison of Tucson and Cracow longitudinal studies, *Chest*, 98:62, 1990.

Krzyzanowski, M., Quackenboss, J. J., and Lebowitz, M. D., Chronic respiratory effects of indoor formaldehyde exposure, *Environ. Res.* 52:117, 1990.

Krzyzanowski, M., Quackenboss, J. J., and Lebowitz, M. D., Sub-chronic respiratory effects from long-term exposure to ozone in Tucson, *Arch. Environ. Health*, 47:107, 1992.

Lambert, P. M. and Reid, D. D., Smoking, air pollution and bronchitis in Britain, *Lancet*, 1:853, 1970.

Lawther, P. J., Compliance with the clean air act: medical aspects, *J. Inst. Fuel*, 36:341, 1963.

Lawther, P. J., Waller, R. E., and Henderson, M., Air pollution and exacerbations of bronchitis, *Thorax*, 25:525, 1970.

Lawther, P. J., Brooks, A. G., Lord, P. W., and Waller, R. E., Day-to-day changes in ventilatory function in relation to the environment, Part I — spirometric values, *Environ. Res.*, 7:24, 1974a.

Lawther, P. J., Brooks, A. G., Lord, P. W., and Waller, R. E., Day-to-day changes in ventilatory function in relation to the environment, Part II — peak expiratory flow values, *Environ. Res.*, 7:41, 1974b.

Lawther, P. J., Brooks, A. G., Lord, P. W., and Waller, R. E., Day-to-day changes in ventilatory function in relation to the environment, Part III — frequent measurements of peak flow, *Environ. Res.*, 8:119, 1974c.

Leaderer, B. P., Berman, M. D., and Stolwijk, J. A. J., *Proceedings of the Fourth International Clean Air Congress*, Kasuga, S. et al., Eds., in *JUAPPA*, Tokyo, pp. 1, 1977.

Leaderer, B. P., Zagraniski, R. T., Berwick, M., and Stolwijk, J. A. J., Assessment of exposure to indoor air contaminants from combustion sources: methodology and application, *Am. J. Epidemiol.*, 124:275, 1986.

Lebowitz, M. D., A comparison of the relationship of mortality and morbidity with air pollution-weather and the implications for further research, *Sci. Total Environ.*, 2:191, 1973a.

Lebowitz, M. D., Toyama, T., and McCarroll, J. R., The relationship between air pollution and weather as stimuli and daily mortality as response in Tokyo, Japan with comparison with other cities, *Environ. Res.*, 6:327, 1973b.

Lebowitz, M. D., Utilization of data from human population studies for setting air quality standards: evaluation of important issues, *Environ. Health Perspect.*, 52:193, 1983.

Lebowitz, M. D., Holberg, C. J., Boyer, B., Hayes, C., Respiratory symptoms and peak flow associated with indoor and outdoor air pollutants in the Southwest, *J. Am. Pollut. Control Assoc.* 35:1154, 1985.

Lebowitz, M. D., and Burrows, B., Risk factors in induction of lung disease: an epidemiologic approach, in *Mechanisms of Lung Injury*, Stickley Publ. Co., Philadelphia, 1986, pp. 208.

Lebowitz, M. D., Efectos agudos de los contaminantes del aire, in *Documentos de Divulgacion*, PAHO (Spanish), 1988a.

Lebowitz, M. D., Sulfate and nitrate aerosols: their effects on human health, in *Documentos de Divulgacion*, Pan American Health Organization, PAHO, Washington D.C., 1988b.

Lebowitz, M. D., and Quackenboss, J. J., The effects of environmental tobacco smoke on pulmonary function, *Int. Arch. Occap. Environ. Health*, Suppl.:147, 1989.

Lebowitz, M. D., Populations at risk: addressing health effects due to complex mixtures with a focus on respiratory effects, *Environ. Health Perspec.*, 95:35, 1991.

Lebowitz, M. D., Indoor bioaerosol contaminants, in *Environmental Toxicants: Human Exposures and Their Health Effects*, Van Nostrand Reinholt, New York, 1991, Chap. 11.

Lebowitz, M. D., Sherrill, D. L., and Holberg, C. J., Effects of passive smoking on lung growth in children, *Ped. Pulmonol.*, 12(1):37, 1992a.

Lebowitz, M. D., Quackenboss, J. J., Krzyzanowski, M., O'Rourke, M. K., and Hayes, C., Multipollutant exposures and health responses: epidemiological aspects of particulate matter, *Arch. Environ. Health*, 47:71, 1992b.

Lebowitz, M. D. and Spinaci, S., The epidemiology of asthma, *Eur. Respir. Rev.*, 3(14):415, 1993.

Lebowitz, M. D., Pulmonary responses to multipollutant airborne particulate matter and other contaminants, with prevention strategies, in *Prevention of Respiratory Diseases*, Marcell Dekker, New York, 1993, pp. 209.

Lebowitz, M. D., Epidemiological and biomedical interpretations of PM10 results: issues and controversies, *Inhal. Toxicol.*, 7(5), 7:757, 1995.

Lioy, P. J., Vollmuth, T. A., and Lippmann, M., Persistence of peak flow decrement in children following ozone exposures exceeding the national ambient air quality standard, *JAPCA*, 35:1068, 1995.

Lippmann, M., Health effects of ozone: a critical review, *J. Air Pollut. Control Assoc.*, 39:672, 1989.

Lowrance, W. W., Of acceptable risk, in *Science and the Determination of Safety*, Kaufman, Los Altos, 1976.

Lunn, J. E., Knowelden, J., and Roe, J. W., Patterns of respiratory illness in Sheffield infant schoolchildren, *Br. J. Prev. Med.*, 21:47, 1967; A follow-up study, *Br. J. Prev. Soc. Med.*, 24:223, 1970.

Molfino, N. A., Wright, S. C., Katz, I., Tarb, S., Silverman, F., McClean, P. A., Szalai, J. P., Raizenne, M., Slutzsky A. S., and Zamel, N., Effect of low concentrations of ozone on inhaled allergen responses in asthmatic subjects, *Lancet*, 338:199, 1991.

Moschandreas, D. J., Exposure to pollutants and daily time budgets of people, *Bull. N.Y. Acad. Med.*, 57:845, 1981.

Moseholm, L., Taudorf, E., and Frosig, A., Pulmonary function changes in asthmatics associated with low-level SO_2 and NO_2 air pollution, weather, and medicine intake — an 8-month prospective study, *Allergy*, 48:334, 1993.

National Academy of Sciences (NAS, NRC Committees): *Indoor Pollutants*, 1981; *Epidemiology*, 1985; *Human Exposure Assessment to Airborne Particulates*, 1991; *Aero-allergens*, 1993. National Academy Press, Washington, D.C.

Neas, L. M., Dockery, D. W., Ware, J. H., Spengler, J. D., Speizer, F. E., and Ferris, Jr., B. G., Association of indoor nitrogen dioxide with respiratory symptoms and pulmonary function in children, *Am. J. Epidemiol.*, 134:204, 1991.

Newman-Taylor, A., and Tee, R. D., Environmental and occupational asthma: exposure assessment, *Chest*, 98(5):209S, 1990.

O'Rourke, M. K. and Lebowitz, M. D., The importance of environmental allergens in the development of allergic and chronic obstructive diseases, in *Environmental Respiratory Diseases*, Van Norstrand Reinholt, New York, (in press), 1995.

Phalen, R. (Ed.), PM10 Effects: a colloquium, *Inhal. Toxicol.*, 75(1&2): 1-392, 1995.

Ponka, A., Asthma and low level air pollution in Helsinki, *Arch. Environ. Health*, 46:262, 1991.

Pope, III, C. A., Dockery, D. W., Spengler, J. D., and Raizenne, M. E. Respiratory health and PM10 pollution: a daily time series analysis, *Am. Rev. Respir. Dis.*, 144:668, 1991.

Pope, C. A., Schwartz, J., and Ransom, M. R., Daily mortality and PM10 Pollution in Utah Valley, *Arch. Environ. Health*, 47:211, 1992.

Quackenboss, J. J., Lebowitz, M. D., Hayes, C., and Young, C. L., Respiratory responses to indoor/outdoor air pollutants: combustion products, formaldehyde, and particulate matter, in *Combustion Processes and the Quality of the Indoor Air Environment*, Harper, J., Ed., AWMA, Pittsburgh, 1989, pp. 280.

Quackenboss, J. J., Krzyzanowski, M., and Lebowitz, M. D., Exposure assessment approaches to evaluate respiratory health effects of particulate matter and nitrogen dioxide, *J. Expos. Anal. Environ. Epidemiol.*, 1:83. 1991.

RIVM, Indoor environment, in *A National Environmental Survey 1985–2010, Concern for Tomorrow*, National Institute of Public Health and Environmental Protection, The Netherlands, 1989, 243.

Robbins, A. S., Abbey, D. E., and Lebowitz, M. D., Passive smoking and chronic respiratory disease symptoms in non-smoking adults, *Int. J. Epidemiol.*, 22:809, 1993.

Roemer, W., Hoek, G., and Brunekreef, B., Effect of ambient winter air pollution on respiratory health of children with chronic respiratory symptoms, *Am. Rev. Respir. Dis.*, 147:118, 1993.

Rombout, P. J. A., Lioy, P. J., and Goldstein, B. D., Rationale for an eight-hour ozone standard, *JAPCA*, 36:913, 1986.

Romieu, I., Cortes Lugo, M., Ruiz Velasco, S., Sanchez, S., Meneses, F., and Hernandez, M., Air pollution and school absenteeism among children in Mexico City, *Am. J. Epidemiol.*, 136:1524, 1992.

Rossi, O. V. J., Kinnula, V. L., Tienari, J., and Huhti, E. Association of severe asthma attacks with weather, pollen, and air pollutants, *Thorax*, 48:244, 1993.

Samet, J., Ed., Environmental controls and lung disease, *Am. Rev. Respir. Dis.*, 142:915, 1990.

Schwartz, J. and Dockery, D. W., Particulate air pollution and daily mortality in Steubenville, Ohio, *Am. J. Epidemiol.*, 135:12, 1992.

Schwartz, J., Slater, D., Larson, T. V., Pierson, W. E., and Koenig, J. Q., Particulate air pollution and hospital emergency room visits for asthma in Seattle, *Am. Rev. Respir. Dis.*, 147:826, 1993.

Seltzer, J., Bigby, B. G., Stulbarg, M., Holtzman, M. J., Nadel, J. A., Ueki, I. F., Leikauf, G. D., Goetzl, E. J., and Boushey, H. A., O_3-induced change in bronchial reactivity to methacholine and airway inflammation in humans, *J. Appl. Physiol.*, 60:1321, 1986.

Sherrill, D., Sears, M. R., Lebowitz, M. D., Holdaway, M. D., Hewitt, C. J., Flannery, E. M., Herbison, G. P., and Silva, P. A., Airway hyperresponsiveness and lung function in children, *Ped. Pulmonol.*, 13:78, 1992.

Sporik, R., Holgate, S. T., Platts-Mills, T. A. E., and Cogswell, J. J., Exposure to house-dust mite allergen (Der pI) and the development of asthma in childhood, *N. Engl. J. Med.*, 323:502, 1990.

Stebbings, J. H., and Fogelman, D. G., Identifying a susceptible subgroup: effects of the Pittsburgh air pollution episode upon schoolchildren, *Am. J. Epidemiol.*, 110:27, 1979.

Sunyer, J., Saez, M., Murillo, C., Castellsavge, J., Martinez, F., and Anto, J. M., Air pollution and emergency room admissions for COPD: a five year study, *Am. J. Epidemiol.*, 137:701, 1993.

Taylor, R. G., Joyce, H., Gross, E., Holland, F., and Pride, N. B., Bronchial reactivity to inhaled histamine and annual rate of decline in FEV1 in male smokers and ex-smokers, *Thorax*, 40:9, 1985.

Trigg, C. J. and Davies, R. J., Infection, asthma and bronchial responsiveness, *Respir. Med.* 87:165, 1993.

US EPA, Air Quality Criteria Documents and Staff Papers, RTP (NC): ECAO and OAQPS, 1982a; Ozone and Other Photochemical Oxidants, V.5, 1984; Particulate Matter and Sulfur Oxides, 1982b; 2nd Addendum, 1986 (b); Summary Report on Atmospheric Nitrates, 1974; Environmental Tobacco Smoke and Respiratory Diseases, 1992.

US NCHS (National Center for Health Statistics), Current Estimates from the U.S. National Health Interview Survey, DHHS Publ. (PHS); Hospital Use in Poland and the U.S., Publ. No. (PHS) 88-1478, 1988.

US NHLBI, Guidelines for the Diagnosis and Management of Asthma, (NIH Pub. No. 91-3042), 1991.

US NIAID, *Asthma and Other Allergic Diseases*, (NIH Pub.No. 79-387), 1979.

Utell, M. J., and Samet, J. M., Particulate air pollution and health — new evidence on an old problem, *Am. Rev. Respir. Dis.*, 147:1334, 1993.

Von Mutius, E., Fritzsch, C., Weiland, S. K., Roll, G., and Magnussen, H., Prevalence of asthma and allergic disorders among children in united Germany: a descriptive comparison, *BMJ*, 305:1395, 1992.

Waller, R. E., and Swan, A. V., Invited commentary: particulate air pollution and daily mortality, *Am. J. Epidemiol.*, 135:1, 1993.

Weiss, K. B., Gergen, P. J., and Hodgson, T. A., An economic evaluation of asthma in the US, *N. Engl. J. Med.*, 326:862, 1992.

Whittmore, A. S. and Korn, E. L., Asthma and air pollution in the Los Angeles area, *Am. J. Public Health*, 70:687, 1980.

WHO, Indoor Air Pollutants: Exposure and Health Effects, Report on a WHO meeting, EURO report and studies n. 78, Copenhagen WHO Regional Office for Europe, 1983.

WHO, Indoor Air Quality: Radon and Formaldehyde, Report on a WHO meeting, Dubrovnik, 26–30 August 1985, WHO regional publications, Environmental Health n. 13, Copenhagen, WHO Regional Office for Europe, 1986a.

WHO, Indoor Air Quality Research, Report on a WHO meeting, Stockholm, 27–31 August 1984, (Document EUR/ICP/RUD 033), Copenhagen, WHO Regional Office for Europe, 1986b.

WHO, Air Quality Guidelines for Europe, Copenhagen WHO Regional Office for Europe, 1987.

WHO, Indoor Air Quality: Organic Pollutants, Report on a WHO meeting, Berlin (West), 23–27 August 1987, (Document EUR/ICP/CEH 026), Copenhagen, WHO Regional Office for Europe, 1989.

WHO, Indoor Air Quality: Biological Contaminants, Report on a WHO meeting, Rautavaara, 29 August–2 September 1988, WHO regional publications European series n. 31, Copenhagen, WHO Regional Office for Europe, 1990.

WHO, Impact on Human Health of Air Pollution in Europe, (Document EUR/ICP/CEH 097–1499n), Copenhagen, WHO Regional Office for Europe, 1991.

WHO, *Concern for Europe's Tomorrow: Summary*, Copenhagen, WHO Regional Office for Europe, 1994.

Zagraniski, R. T., Leaderer, B. P., and Stolwijk, J. A., Ambient sulfates, photochemical oxidants, and acute health effects: an epidemiological study, *Environ. Res.*, 19:306, 1979.

Zwick, H., Popp, W., Wagner, C., Reiser, K., Schmoger, J., Bock, A., Herkner, K., and Radunsky, K., Effects of ozone on the respiratory health, allergic sensitization, and cellular immune system in children. *Am. Rev. Respir. Dis.*, 144:1075, 1991.

PART 3: METHODS

12   SMALL-AREA STUDIES

Paul Elliott

CONTENTS

INTRODUCTION

Geographical studies have a long tradition in epidemiology. International, and, to a lesser extent, national studies are able to exploit large geographical variation both in disease rates and in exposure to environmental and lifestyle factors such as smoking and diet. For example, large international differences in the occurrence of heart disease (Keys, 1970) or colon cancer (Armstrong and Doll, 1975) have led to hypotheses as to their aetiology, which have then been explored at the individual level. In general, the use of such broad scale, ecological studies in establishing aetiology is often limited by the problem of unmeasured confounding. This can seriously bias ecological relationships and lead to misleading or false conclusions concerning relationships at the individual level. (Elliott, 1992; English, 1992).

More recently, there has been considerable interest in exploring much more local geographical variations in disease, often in response to public or media concern about a potential (or perceived) environmental hazard. In the U.K., scarcely a week goes by without a television or newspaper report of a putative disease cluster, although informed interpretation of these reports is often difficult if not impossible using the published information (Dolk and Elliott, 1993). It has been stressed that such reports rarely lead to advances in our understanding of disease (Rothman, 1990); nevertheless, local public health authorities are compelled to respond, either to allay public anxiety if no disease excess is found, or to set in train appropriate further investigation if an important excess of disease is confirmed.

Even an initial check on the validity of the data leading to such a report may prove difficult, and a preliminary assessment of the size and extent of any excess risk in comparison with the expected numbers of cases may not be possible. Reports of local disease clusters are unlikely to respect the administrative boundaries that traditionally have determined the reporting of health statistics; and, in any case, a disease excess may occur on a much finer geographical scale and hence go undetected in the routine statistics. Under such circumstances, little progress can be made in the further investigation of a putative cluster without embarking on a detailed (and costly) epidemiological enquiry, unless a routine system for the assembly, storage, and retrieval of small-area geographically referenced data is in place.

In the U.K., one such media report of an alleged increase of childhood leukaemia near the Sellafield nuclear reprocessing plant led to a series of epidemiological studies which continue to this date, some 10 years later. These studies have suggested a possible new aetiology for childhood leukaemia related to occupational exposure of the fathers to ionising radiation around the time of conception (Gardner et al., 1990). More generally, the Sellafield enquiry generated considerable scientific interest in the statistical and epidemiological methods available for small-area studies, and led directly to the setting up of the Small Area Health Statistics Unit (SAHSU) at the London School of Hygiene and Tropical Medicine.

In the remainder of this chapter, I shall offer a definition of a small-area study and briefly review the different types of small-area epidemiological enquiry. The background and methods of SAHSU are then presented, including discussion of the problem of socio-economic confounding in small-area studies. Finally, some priorities for future research are discussed.

WHAT IS A SMALL-AREA STUDY?

One definition of a small-area study is an analysis carried out at sub-national or perhaps sub-regional level. This definition is not helpful, as the population size in the geographical units may differ, even within one country, by many thousands, tens of thousands, or even hundreds of thousands of people. A more useful definition would be determined by the number of cases, which itself would depend on population size, disease frequency, and time-period of analysis. As a rough guide, any region containing fewer than about 20 cases of disease can be considered a small area. Many cancers have annual incidence rates of around 5 per 100,000, so that over a 5-year period a small area might comprise a population of 100,000 or less. For rare diseases or small populations in remote areas, the population size might be much less, but usually populations of at least 10,000 or so are required to form an aggregation of minimal size (Cuzick and Elliott, 1992).

TYPES OF SMALL-AREA STUDIES

In a recent review, Cuzick and Elliott (1992) identified seven types of small-area study, as follows:

Studies of reports of disease excess ("clusters") in specific localities without a putative source — In this type of study, problems of interpretation are severe as many apparent clusters are certain to arise by chance. Apparent clusters, especially of rare diseases, are likely to come to public attention before they are known to the health authorities. The clusters may subsequently be linked to a putative source, for example, the excess of childhood leukaemias observed in Woburn, Massachusetts which was linked to contaminated well water (Lagakos et al., 1986).

Studies of point sources of industrial pollution — Public concern about an industrial source of pollution is commonly encountered by local departments of health and environment, and may lead to a review of the available health statistics for the area in the vicinity of the plant. Where appropriate data are readily available, a major problem in interpretation is how to deal with *post hoc* reports of disease excess near an industrial source, since associations that are made in a *post hoc* fashion invalidate standard statistical testing. One approach, if the data systems are in place, is to replicate the study around other similar industrial sites (if such can be found) based on a hypothesis that is driven by the original observation. An example of such an approach can be found in a recent study of cancer incidence near industrial incinerators in Great Britain, which followed reports of an excess of cancer of the larynx near one of the sites (Elliott et al., 1992a, and summarised below).

Studies of clustering as a general phenomenon — The purpose here is to examine for some general tendency of a disease to show patterns of clustering, which might give clues as to aetiology. One example is that of

Hodgkin's disease (Alexander et al., 1989) where the tendency for local clustering has led to suggestions of an infectious aetiology. However, other explanations are possible including data artefacts related, in particular, to problems with the denominator data (Elliott et al., 1991, Besag et al., 1991).

Ecological-correlation studies of health and the environment — Examples of ecological studies at the small-area level are rare, as often the health, environmental, and population data are not available at the same level of geographical resolution. This is, however, a potentially valuable and relatively unexplored approach in environmental epidemiology, and requires further development and evaluation. The problems of interpretation of ecological studies mentioned in the introduction may be less severe at this geographical scale, as the unit of analysis is closer to the individual.

Descriptive studies at the small-area level — Little is known of the natural variation of disease rates over small areas, although such information is necessary in order to place specific disease clusters in proper context. An example of this type of study is the examination of the geographical epidemiology of childhood leukaemia in Great Britain (Draper, 1991), in which small-area variations in incidence were found to be related both to regional and to socioeconomic factors (Rodrigues et al., 1991).

Geographical surveillance — Computer systems for carrying out small-area analyses of routine data offer the possibility of undertaking automatic surveillance in order to detect areas of high disease incidence (Openshaw and Craft, 1991). However, many false positives are bound to occur using currently available methods. Until such time as an acceptable theoretical framework for such analyses is in place, these endeavours need to be viewed as experimental in nature and any substantive results regarded with extreme caution.

Acute accidents — Studies of acute chemical or nuclear accidents have much in common with studies of point sources of pollution, with the added dimension of time. An example is the explosion at the chemical plant in Seveso, where geographically defined cohorts were identified according to presumed exposure to dioxins (Bertazzi et al., 1992).

THE SMALL-AREA HEALTH STATISTICS UNIT

SAHSU is an independent national facility, funded by U.K. government, for the investigation of disease near point sources of industrial pollution (Elliott et al., 1992b, c). Its terms of reference are as follows:

1. To examine quickly reports of unusual clusters of disease, particularly in the neighbourhood of industrial installations and advise authoritatively as soon as possible
2. In collaboration with other scientific groups, to build up reliable background information on the distribution of disease among small areas so that specific clusters can be placed in proper context

3. To study the available statistics in order to detect any unusual incidence of disease as early as possible and, where appropriate, to investigate
4. To develop the methodology for analysing and interpreting statistics relating to small areas

We have been particularly concerned with the study of disease around multiple industrial sites, either to replicate an enquiry conducted *post hoc* around one site (by studying other sites in Britain producing similar discharges) or to test hypotheses related to a particular industrial process. Current projects include the investigation of cancer incidence near municipal incinerators and radio transmitters, and the incidence of angiosarcoma of the liver near vinyl chloride plants.

SAHSU incorporates a national database which includes mortality (by specific cause), cancer registration data, and congenital malformation data. Data retrieval is based on the postcode of residence at the time of death or diagnosis. The postcode relates to only 14 households on average, and can be located as a point on a map to 100 m accuracy. Population data for small areas (enumeration districts) are available from national census. A typical enumeration district gives population counts and socio-economic data for about 400 people. By use of a database retrieval system with postcode as a key, rates of mortality and cancer incidence can rapidly be analysed for arbitrary circles located anywhere in Britain. Recent addition of a Geographical Information System (GIS) has meant that occurrence of disease in relation to irregularly shaped areas (e.g., wind pattern) or linear structures (e.g., coastline, roads, power lines) can also be readily determined.

Postcoded data sets held by SAHSU include:

1. Deaths for England and Wales from 1981, for Scotland from 1974, and for Northern Ireland from 1986
2. Live- and stillbirths for England and Wales from 1981, the birth data providing accurate year-by-year postcoded denominator counts for perinatal and childhood events
3. Cancer registrations for England and Wales from 1974 and Scotland from 1975; enhanced data on childhood cancers in Great Britain being supplied by the Childhood Cancer Research Group in Oxford
4. Congenital malformations for England and Wales from 1983

For each individual event, SAHSU holds postcode of residence, diagnosis (ICD [International Classification of Disease] code, and histology code for cancer registrations), age in years and months, and identifier allowing linkage to individual records through the Office of Population, Censuses and Surveys (OPCS). A similar system is in place for Scottish data.

The SAHSU system runs under UNIX using reduced instruction set computer (RISC) technology including a Digital (DEC) 5500 super-microcomputer and a Sun Sparcserver 10 Model 41 with dual processor. There are around

15,000 Mbytes of data storage available. We use the Oracle relational database management system and implement algorithms in C with embedded SQL for access to the database. Geographical data retrieval and mapping are done using Arc-Info, with links to the Oracle database. Statistical analysis is done mainly in the statistical package Splus.

For the analyses around a point source, currently only a simple radial dispersion model is used, although extension to other models (to take account, for example, of prevailing wind patterns) is feasible using the GIS as mentioned above. Thus a range of circle sizes around a source is chosen *a priori*; and the numbers of events observed, and the numbers expected, are calculated for the bands between adjacent circles. Expected numbers are calculated from national rates standardized for age and sex. Adjustment is also made for a measure of the socio-economic profile of small areas, and for region — to allow for regional variation in disease incidence and (for cancers) for variation in the quality and completeness of registration.

Currently, data are examined using two methods. The first is based on the Poisson distribution, to give observed/expected (O/E) ratios and 95% confidence intervals for two adjacent bands close to, and more distant from, the source. The second uses an adaptation of Stone's (1988) method for data over a range of circles, to test whether there is evidence of (1) a general excess risk above expected, and (2) decreasing risk with distance from the source. Data from a number of installations can readily be pooled and tested in the same way.

An Example: Cancer of the Larynx and Lung Near Incinerators of Waste Solvents and Oils in Great Britain

A report of cancer incidence near a defunct incinerator of waste solvents and oils at Charnock Richard, Coppull, Lancashire, northwest England, suggested that there was an apparent cluster of five cases of cancer of the larynx nearby. This *post hoc* finding was examined using the SAHSU database, and the enquiry extended to all ten eligible incinerator sites in Great Britain that could be identified as burning a similar type of waste. Using Stone's method, the excess of cancer of the larynx near Charnock Richard was found to be within chance limits, based on a small number of cases. In an analysis pooled over all ten sites (which ensured adequate statistical power) no statistically significant excess of either cancer was found. It was concluded that the apparent cluster of cancer of the larynx previously observed at Charnock Richard was unlikely to be due to its former incinerator (Elliott et al., 1992a).

The Problem of Socio-Economic Confounding

A major problem in the interpretation of analyses of disease rates in small areas is the issue of socio-economic confounding (Jolley et al., 1992). The problem is illustrated in the figures. Figures 1 and 2 show, for males and females, the standardised incidence ratio in 1981 of a number of cancers according to a small-area deprivation measure based on 1981 census statistics, for the 130,000 enumeration districts across Great Britain. The enumeration districts have been ranked into quintiles according to their deprivation score — the most deprived areas are in the highest quintile. A sixth (unclassified) group is included for the 5% of enumeration districts where small numbers in one of the cells preclude an assignment of a deprivation score. Figure 3 shows the relationship of deprivation score with distance from an industrial source.

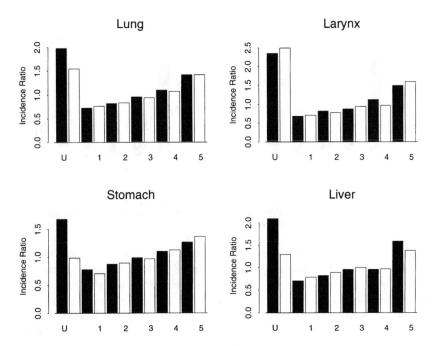

Figure 1   Standardised incidence ratios for cancers of the lung, larynx, stomach, and liver in relation to quintiles of deprivation scores for enumeration districts, Great Britain, 1981. Areas of greater deprivation are in the higher quintiles. Data were unclassified (U) for 5% of enumeration districts. Data for men are shown by the black boxes and for women by the open boxes.

The census variables used in the calculation of the deprivation scores are persons in households with more than one person per room; persons in households where the head is economically active and from social class IV or V; and economically active males seeking work. This is similar to the index of deprivation described by Carstairs (1990) except that the Carstairs index includes in addition "persons in private households without access to a car". The two measures are highly correlated (Spearman rank correlation coefficient

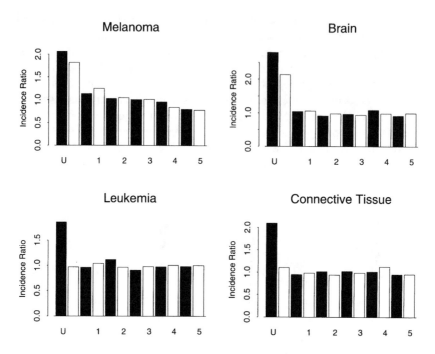

Figure 2   Standardised incidence ratios for melanoma, brain cancer, leukemia, and cancers of connective tissue, in relation to quintiles of deprivation scores for enumeration districts, Great Britain, 1981. For key see Figure 1.

rho = 0.97), but with the three-variable deprivation score used here slightly fewer of the enumeration districts are included in the unclassified band.

Figure 1 shows data for four cancers with strong positive gradients in incidence across the deprivation quintiles (i.e., higher incidence in the more deprived areas). For both sexes combined, the ratio of Q5/Q1 is 1.92 for lung, 2.21 for larynx, 1.75 for stomach, and 2.04 for liver. The highest standardised incidence ratios are recorded for the small unclassified band. This contains enumeration districts that are predominated, for example, by old people's homes or nursing homes. Figure 2 shows data for cancers where the deprivation gradient is in the reverse direction (melanoma) or is essentially flat (brain, leukaemia, connective tissue). In Figure 3, the deprivation score is shown for each of the enumeration districts whose centroids (represented by a point) lie within 10 km of an industrial source — an incinerator in this example. The more deprived areas are again represented by higher scores. The continuous line shows the smoothed median deprivation score with distance. As can be seen, the median score declines rapidly over 10 km from the source, although there is a small blip at around 5 km.

Taken together, interpretation of the pattern of, say, lung cancer incidence near to the source shown in Figure 3, is complicated by the fact that any excess

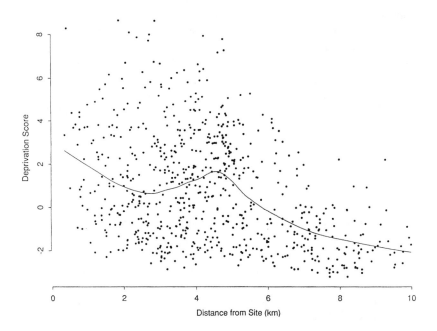

Figure 3   Deprivation scores of enumeration districts (indicated by points) in relation to distance from a point source. The line indicates smoothed median deprivation score with distance.

could be due entirely to confounding by socioeconomic factors, which are themselves likely to be related to smoking rates. In SAHSU, we therefore routinely take socioeconomic factors into account in the calculation of disease risks around a point source. Enumeration districts are grouped into five equal strata according to deprivation score, as shown in the figures, as well as the small sixth stratum for unclassified areas. National stratum-age-sex specific rates are then applied to the underlying populations to calculate stratified expected values, which are then summed up in the usual way. Ratios of expected values stratified/unstratified by the deprivation index can be examined to give an indication of the extent of possible socio-economic confounding.

SOME PRIORITIES FOR FUTURE RESEARCH

Availability of Data and Confidentiality

Few countries except for the U.K. and the Scandinavian countries have the capability of carrying out small-area analyses using routine data. A major constraint is the availability of appropriate health data at the small-area level, such as cancer registrations, often because of reasons of data confidentiality; indeed, the proposed European Community (EC) directive on confidentiality could have a devastating effect on the ability of bona fide researchers to use national or regional registries for epidemiological purposes. A particular problem in the U.K. is the lack of reliable morbidity data (e.g., hospital discharge data) other than cancer registrations, to allow for the analysis of acute effects such as respiratory admissions. A further constraint in many countries is the lack of a system of geo-referencing of health and population data (similar to the use of the postcode in the U.K.). Until these problems are overcome, it will not be possible for most countries to respond rapidly to the need for small-area enquiries.

Exposure Data

Environmental pollution data tend to be collected at a small number of fixed monitoring sites across a country. Much more needs to be known about the variation in the common pollutants over small geographical areas. Although stack emission sampling may be carried out at an industrial site, the collection of environmental measurements around a site and environmental modelling are rarely done. In the absence of such data, small-area epidemiological studies are bound to use proxy measures of true exposure, such as distance from a point source, with the inevitable problems of misclassification of exposure and underestimation of risk.

Socio-Economic Confounding

Our preliminary analyses suggest that even with correction for differences in small-area deprivation as indicated above, important residual confounding by socio-economic factors is likely to occur in some areas. Further research is needed to identify improved methods for dealing with socio-economic confounding at the small-area level; the application of proportional morbidity or mortality measures may prove useful in this regard. Research is needed also to compare these ecological measures with relevant aetiological factors at the individual level, e.g., smoking and alcohol consumption.

Matching Health, Population, and Environment Data

There is an urgent need for appropriate methodology to combine health, population, and environment data at the same (small-area) level of geographical resolution; similarly, appropriate methods of statistical analysis for those data need to be developed and applied. As an example, the EC is funding a small-area methodological study (Small Area Variations in Air quality and Health — SAVIAH) based in the U.K., The Netherlands, the Czech Republic, and Poland; and coordinated at the London School of Hygiene and Tropical Medicine. The study is investigating the use of passive samplers for the measurement of NO_2 and SO_2 in small areas, the use of GIS and pollution modelling techniques, and the use of hierarchical random effects models in the statistical analysis of outdoor air pollution in relation to childhood wheezing and asthma.

SUMMARY

The study of the effects of environmental pollution on health is complicated by problems of data quality, and the geographical resolution at which relevant data are available. Recent advances in methods for small-area studies have meant that the initial investigation of disease near point sources of pollution can largely be automated, although care is needed in the interpretation of a positive result, especially where there is the possibility of substantial socio-economic confounding. More widespread use of small-area methods, to enable the rapid replication of findings internationally as well as nationally, would enhance our ability to quantify the effects of environmental pollution on human health.

ACKNOWLEDGMENTS

The Small Area Health Statistics Unit (SAHSU) is funded by grants from the Department of Health, Department of the Environment, Health and Safety Executive, Scottish Office Home and Health Department, Welsh Office, and the Northern Ireland Department of Health and Social Services. I am grateful to Immo Kleinschmidt and Peter Walls (London School of Hygiene and Tropical Medicine) for preparing the figures.

REFERENCES

Alexander, F. E., Williams, J., McKinney, P. A., Cartwright, R. A., and Ricketts, T. J. (1989). A specialist leukaemia/lymphoma registry in the UK. Part 2: clustering of Hodgkin's disease. *Br. J. Cancer* 60:948-952.

Armstrong, B. K. and Doll, R. (1975). Environmental factors and cancer incidence and mortality in different countries, with special reference to dietary practices. *Int. J. Cancer* 15:617-631.

Besag, J., Newell, J., and Craft, A. (1991). The detection of small-area anomalies in the database. In: *The Geographical Epidemiology of Childhood Leukaemia and Non-Hodgkin Lymphomas in Great Britain, 1966-83*, Draper, G., Ed., Studies in Medical and Population Subjects No. 53. London: HMSO, chap. 11, pp. 101-107.

Bertazzi, P. A., Pesatori, A. C., and Zocchetti, C. (1992). The Seveso accident. In: *Geographical and Environmental Epidemiology: Methods for Small Area Studies*, Elliott, P., Cuzick, J., English, D., Stern, R., Eds., Oxford: Oxford University Press, pp. 342-358.

Carstairs, V. (1990). Deprivation and health in Scotland. *Health Bull.* 48:162-175.

Cuzick, J. and Elliott, P. (1992). Small area studies: purpose and methods. In: *Geographical and Environmental Epidemiology: Methods for Small Area Studies*, Elliott, P., Cuzick, J., English, D., Stern, R., Eds., Oxford: Oxford University Press, pp. 14-21.

Dolk, H. and Elliott, P. (1993). Evidence for "clusters of anophthalmia" is thin. *Br. Med. J.* 307: 203.

Draper, G., Ed, (1991). *The Geographical Epidemiology of Childhood Leukaemia and Non-Hodgkin Lymphomas in Great Britain, 1966-83*. Studies in Medical and Population Subjects No. 53. London: HMSO, pp. 17-23.

Elliott, P. Design and analysis of multicentre epidemiological studies: the INTERSALT Study. In: *Coronary Heart Disease Epidemiology: from Aetiology to Public Health*, Marmot, M., and Elliott, P., Eds., Oxford: Oxford University Press, 1992, chap. 12, pp. 166-178.

Elliott, P., McGale, P., and Vincent, T. J. (1991). Description of population data and definition of areas. In: *The Geographical Epidemiology of Childhood Leukaemia and Non-Hodgkin Lymphomas in Great Britain, 1966-83*, Draper, G., ed., Studies in Medical and Population Subjects No. 53. London: HMSO, chap. 3, pp. 17-23.

Elliott, P., Hills, M., Beresford, J., Kleinschmidt, I., Jolley, D., Pattenden, S., Rodrigues, L., Westlake, A., and Rose, G. (1992a). Incidence of cancer of the larynx and lung near incinerators of waste solvents and oils in Great Britain. *Lancet* 339: 854-858.

Elliott, P., Westlake, A. J., Kleinschmidt, I., Rodrigues, L., Hills, M., McGale, P., Marshall, K., and Rose, G. (1992b). The Small Area Health Statistics Unit: a national facility for investigating health around point sources of environmental pollution in the United Kingdom. *J. Epidemiol. Commun. Health* 46:345-349.

Elliott, P., Kleinschmidt, I., and Westlake, A. J. (1992c). Use of routine data in studies of point sources of environmental pollution. In: *Geographical and Environmental Epidemiology: Methods for Small Area Studies*, Elliott, P., Cuzick, J., English, D., and Stern, R., Eds., Oxford: Oxford University Press, pp. 106-114.

English, D. (1992). Geographical epidemiology and ecological studies. In: *Geographical and Environmental Epidemiology: Methods for Small Area Studies*, Elliott, P., Cuzick, J., English, D., Stern, R., Eds., Oxford: Oxford University Press, pp. 3-13.

Gardner, M. J., Snee, M. P., Hall, A. J., Powell, C. A., Downes, S., and Terrell, J. D. (1990). Results of case-control study of leukaemia and lymphoma among young people near Sellafield nuclear plant in West Cumbria. *Br. Med. J.* 300:423-429.

Jolley, D., Jarman, B., and Elliott, P. (1992). Socio-economic confounding. In: *Geographical and Environmental Epidemiology: Methods for Small Area Studies*, Elliott, P., Cuzick, J., English, D., and Stern, R., Eds., Oxford: Oxford University Press, pp. 115-124.

Keys, A., Ed (1970). *Coronary Heart Disease in Seven Countries*. Am. Heart Assoc. Monogr. no. 29. New York: The American Heart Association.

Lagakos, S. W., Wesser, B. J., and Zelen, M. (1986). An analysis of contaminated well water and health effects in Woburn, Massachusetts (with discussion). *J. Am. Stat. Assoc.* 81:583-614.

Openshaw, S. and Craft, A. (1991). Using geographical analysis machines to search for evidence of clusters and clustering in childhood leukaemia and non-Hodgkin lymphomas in Britain. In: *The Geographical Epidemiology of Childhood Leukaemia and Non-Hodgkin Lymphomas in Great Britain, 1966-83*, Draper, G., Ed., Studies in Medical and Population Subjects No. 53. London: HMSO, chap. 12, pp. 109-122.

Rodrigues, L., Hills, M., McGale, P., and Elliott, P. (1991). Socioeconomic factors in relation to childhood leukaemia and non-Hodgkin lymphomas: an analysis based on small area statistics for census tracts. In: *The Geographical Epidemiology of Childhood Leukaemia and Non-Hodgkin Lymphomas in Great Britain, 1966-83*, Draper, G., Ed., Studies in Medical and Population Subjects No. 53. London: HMSO, pp. 47-56.

Rothman, K. J., (1990). A sobering start for the cluster busters' conference. *Am. J. Epidemiol.* 132 (Suppl. 1):S6-S13.

Stone, R. A. (1988). Investigations of excess environmental risks around putative point sources: statistical problems and a proposed test. *Stat. Med.* 7:649-660.

13 USES OF BIOCHEMICAL MARKERS IN ENVIRONMENTAL EPIDEMIOLOGY

Paolo Vineis

CONTENTS

INTRODUCTION

Biomarkers used in epidemiology are usually included in three general categories: markers of internal dose, markers of early response, and markers of susceptibility. Examples are DNA adducts (internal dose), oncogene mutations or chromosome translocations (putative early response), and metabolic polymorphism (susceptibility). Conventional epidemiology has been a sort of black box discipline until now, since it studied the relationship between exposure and disease (the two extremes of the causal chain), without considering the intermediate mechanistic steps. In addition, it relied often on proxy information, in particular, for exposure assessment. This approach has been extremely useful: in fact, all the known causes of several chronic diseases, including cancer, in humans have been discovered with the tools of conventional epidemiology. Limitations of the conventional approach have become manifest in recent years, when epidemiologists have started to face more complex problems. Exposure levels, at least in western countries, are generally low (e.g., to atmospheric pollution), and the distinction between the exposed and the unexposed population is far from being an easy task. In this context, the tools of conventional epidemiology have a decreased sensitivity due, for example, to misclassification of exposure or to modest increases in risk.

RATIONALE FOR THE USE OF BIOMARKERS

The rationale for the introduction of biomarkers in epidemiology has not been fully discussed. By some colleagues, those who practice molecular epidemiology have been compared to supporters of the germ theory of disease for the end of the twentieth century: they look for subtle mechanistic changes taking place in cancer and other diseases, but overlook external, complex, and socially determined causes ("... cause in epidemiology is not a property of agents, but one of complex systems in which the population phenomena of health and disease occur".[1,2])

Although this worry is understandable, there are situations in which the use of biomarkers is justified. The following is a list of potential uses in the context of environmental epidemiology:

1. Exposure assessment when traditional epidemiologic tools are insufficient (particularly for low levels of exposure and low risks)
2. Multiple exposures or mixtures, when the aim is to disentangle the etiologic role of single agents
3. Estimation of the total burden of exposure to chemicals having the same mechanistic target
4. Investigation of pathogenetic mechanisms
5. Study of individual susceptibility (e.g., metabolic polymorphism, DNA repair)
6. Introduction of putative early response markers in epidemiologic studies

For some of these goals the technology is still developing, and there are only limited applications. For instance, as a marker of integrated exposure (point 3 above) 7-methylguanine has been proposed,[3] but field research is very limited. In other cases field applications have been more extensively explored.

METHODOLOGICAL PROBLEMS

I will consider some open methodological aspects in the use of biomarkers.

1. Reliability data (reproducibility and repeatability) have not been published for most of the markers. Some examples of validation, however, are available,[4] showing that the reproducibility of measurements is far from being complete.

2. The quality of the epidemiologic design is not always acceptable, partly because applications are still made in the context of pilot studies and measurements are expensive.

3. The sample size is often insufficient to have reasonably stable statistical estimates.

4. Surrogate cells or organs, instead of the target organ, are often considered for measurement. The significance of SCEs (Sister Chromatid Exchanges) or DNA adducts in lymphocytes must be more clearly assessed before making direct inferences on the target organ. Metabolic competence is known to vary widely in different organs.

5. An important question is, where does the marker lie in the putative causal chain between exposure and disease? For example, a very specific mutation in the p53 tumour suppressor gene has been reported from patients with hepatocellular carcinoma in China and in South Africa, in areas with endemic exposure to aflatoxin B1.[5,6] The mutation was a G to T transversion in codon 249, i.e., the type of mutation known to be induced by aflatoxin B1 *in vitro*. Although the specificity of the observation is highly suggestive, the two studies are of the cross-sectional type, so that one cannot rule out that the p53 mutation is a side effect of exposure, or some epiphenomenon of cancer, and not a mechanistic step in the causal chain between exposure and cancer.

6. The biological significance of some early response markers is still uncertain. SCEs and chromosome aberrations belong to this category, although some evidence is now becoming available on their predictivity of cancer risks. A follow-up study is ongoing in the Nordic countries: a preliminary report has been published, showing that among 2969 subjects with some form of chromosome alteration, those in the highest tertile for frequency of the alterations had an 80% increase of cancer mortality (all sites).[7] Although this finding is still limited and not statistically significant, it is nevertheless encouraging.

7. A final limitation of current biomarkers, in this partial list, is represented by time. Most markers of internal dose either have a very short half-life, or are very unstable biologically, so that their use in epidemiologic studies of cancer is limited. For example, DNA adducts in lymphocytes represent exposures that took place months before and therefore are not suitable for case-control studies of chronic diseases, in which one aims at elucidating exposures that occurred years in advance. To overcome this problem, the introduction of a nested design (case-control in the cohort) has been proposed; however, costs are considerable, and the stability of markers is not necessarily so high as to make storing of biological material for years a reliable practice.

Methodological aspects of the use of biomarkers in epidemiology are discussed elsewhere.[8]

ESTIMATING THE CONTRIBUTION OF A BIOMARKER

A biomarker does not necessarily perform better than a traditional tool such as a questionnaire. For example, on most occasions a smoking questionnaire is quite sufficient to achieve credible information on smoking habits. The additional contribution of biomarkers can be measured following the logic of clinical epidemiology when evaluating the contribution of a diagnostic test. For example, we know that cotinine and nicotine in the urine tend to be more accurate than the use of a questionnaire; but how much and how cost-effectively? Let us suppose that the positive predictive value of a questionnaire is 0.94: what is the additional gain allowed by cotinine-nicotine measurement? According to clinical epidemiology, the rough evaluation based on the questionnaire is the pre-test probability that the individual smokes, and the cotinine-nicotine measurement represents the post-test probability. The contribution of the biomarker can be therefore estimated as the difference between the two probabilities. One property of this clinical reasoning is that the performance of a test is best in intermediate situations between very low *a priori* chances (say, pre-test probabilities of 0.1–0.2 or less), and situations of almost certainty (probabilities of 0.9 or more). In other words, it would be justifiable to use a biomarker when it can add significantly to previous knowledge, i.e., in conditions of real uncertainty.

In fact, it is not unrealistic to think that a conventional tool sometimes performs better than a biochemical measurement. This can happen when the marker is very specific and applies only to a subgroup of the exposed subjects. For example, the measurement of adducts formed by one nitroarene might be less informative than a good questionnaire on air pollution from diesel exhaust, for at least two reasons:

1. The lower prevalence of subjects with measurable adduct levels, compared to those exposed to diesel exhaust, entails lower predictive value for the same sensitivity and specificity levels.
2. The presence of that specific adduct may reflect peculiar exposure conditions, not representative of a range of qualitatively and quantitatively different situations. This example just indicates that the reasons for the introduction of a marker should be well defined *a priori*.

ETHICAL ASPECTS

Apart from the preceding considerations on the rationale for the use of biomarkers, one should also consider ethical and political aspects. The use of biomarkers in environmental epidemiology must be accompanied by a clear policy as far as informed consent is concerned. A subject may have several reasons to refuse co-operation. One, very practical, is that after the identification of, say, an alteration in an early response marker or of metabolic susceptibility, private insurers might shun this person because he may be more prone to disease.

AN EXAMPLE: THE ROLE OF GENETICALLY BASED METABOLIC POLYMORPHISM IN MODULATING THE EFFECTS OF LOW-LEVEL EXPOSURE

I will discuss briefly an example based on the coupling of a biomarker of internal dose and a biomarker of susceptibility. This example has important public health implications.

The metabolic activation or deactivation of carcinogens is known to vary in human populations. Genetically based metabolic polymorphisms for the activation and deactivation of bladder carcinogens, such as 4-aminobiphenyl (ABP) (present in tobacco smoke and in the general environment at low concentrations), have been extensively studied: in particular, N-acetylation is involved in the detoxification of carcinogenic arylamines. In one study (Table 1), we determined 4-aminobiphenyl-hemoglobin adduct levels in 97 volunteers, along with the N-acetylation (NAT) phenotype and the levels of

Table 1   Distribution of 97 volunteers by levels of 4-aminobiphenyl-hemoglobin adducts (pg/g hemoglobin), by nicotine + cotinine in the 24-hour urine, and by acetylator phenotype

Nicotine + Cotinine in the Urine	Whole group		Acetylator Phenotype		Percent Increase in Slow
	Median	Mean	Fast	Slow	
			(Median)		
0 (n = 44)	22.5	28.5	13	27	107
<1.5 (n = 25)	65	79.5	52	99	90
1.5–2.4 (n = 9)	128	123.7	92	132	43
2.5 + (n = 19)	125	135.3	128	119	−7

Note: Slow acetylators are those with an AFMU/1-methylxanthine ratio lower than 0.6.

Data from Vineis, P. et al. Nature 369, 154–156, 1994.

nicotine and cotinine in the urine.[9] The NAT genotype was determined in 41 subjects. An effect of the slow acetylator phenotype which differed by nicotine and cotinine levels was observed. Among the slow acetylators, 4-ABP adducts increased by 107, 90, 43, and <0% with increasing levels of nicotine-cotinine, compared to fast acetylators. Also among subjects with the slow NAT genotype, higher levels of hemoglobin adducts were found at low-level 4-ABP exposure. DNA adducts in exfoliated bladder cells were more frequent in slow than in fast acetylators. These data suggest that the clearance of carcinogens is slower in the genetically based slow acetylator phenotype, and that this metabolic polymorphism may be particularly influential in modulating the carcinogenic risk at low levels of exposure. Such observations are relevant to risk assessment procedures, particularly in environmental epidemiology.

REFERENCES

1. Vandenbroucke, J. P., Is 'The Causes of Cancer' a Miasma Theory for the End of Twentieth Century? *Int. J. Epidemiol.* 1988; 17: 708-709.
2. Loomis, D. and Wing, S., Is Molecular Epidemiology a Germ Theory for the End of the Twentieth Century? *Int. J. Epidemiol.* 1990; 19: 1-3.
3. Shuker, D. E. G., Determination of N7-methylguanine by immunoassay. In: Bartsch, H., Hemminki, K., and O'Neill, I. K. (Eds.): Methods for detecting DNA damaging agents in humans: applications in cancer epidemiology and prevention. IARC Scientific Publication No. 89, IARC, Lyon, France, 1988.
4. Wogan, G. N., Detection of DNA damage in studies on cancer etiology and prevention. In: Bartsch, H., Hemminki, K., and O'Neill, I. K. (Eds.): Methods for detecting DNA damaging agents in humans: applications in cancer epidemiology and prevention. IARC Scientific Publication No. 89, IARC, Lyon, France, 1988.
5. Hsu, I. C., Metcalf, R. A., Sun, T., Welsh, J. A., Wang, N. J., and Harris, C. C., Mutational hotspot in the p53 gene in human hepatocellular carcinomas. *Nature (London),* 1991; 350: 427-428.
6. Bressac, B., Kew, M., Wands, J., and Ozturk, M., Selective G to T mutations of p53 gene in hepatocellular carcinoma from southern Africa. *Nature (London),* 1991; 350: 429-431.
7. Nordic Study Group on the health risk of chromosome damage. An inter-nordic prospective study on cytogenetic endpoints and cancer risk. *Cancer Genet. Cytogenet.* 1990; 45: 85-92.
8. Vineis, P., Schulte, P., and Voigt, J., Technical variability in laboratory data. In: Schulte, P. and Perera, F. (Eds.): *Molecular Epidemiology.* Academic Press 1993.
9. Vineis, P., Bartsch, H., Caporaso, N., Harrington, A. M., Kadlubar, F. F., Landi, M. T., Malaveille, C., Shields, P. G., Skipper, P., Talaska, G., Tannenbaum, S. R. Genetically based *N*-Acetyltransferase metabolic polymorphism and low-level environmental exposure to carcinogens. *Nature (London),* 1994; 369:154-156.

14   EXPOSURE ASSESSMENT IN ENVIRONMENTAL EPIDEMIOLOGY

Bert Brunekreef

CONTENTS

INTRODUCTION

Human *exposure assessment* is rapidly evolving into a field of its own, as exemplified by the recent foundation of the International Society for Exposure Analysis. Environmental *exposure* can be defined as any contact between a pollutant present in an environmental medium (such as air, water, or food) and a surface of the human body (such as the skin, the gut, or the respiratory tract epithelium).[1] This definition of exposure sets exposure apart from *concentration* on the one hand and *dose* on the other hand. A concentration of a pollutant in some environmental medium refers to its presence, expressed in quantitative terms, in that medium, but there is no human exposure to that pollutant unless there is physical contact between the medium and (body surfaces of) human beings. This distinction is particularly important for environmental media such as soil, which may be polluted, but not necessarily in contact with the human body. A pollution dose, on the other hand, refers to the amount of pollution that actually crosses the border between the environment and the human body. After the pollutant has entered the human body, toxicokinetics determine how it will be distributed, and which fraction of it reaches the *target* tissue or organ where it may exert some harmful effect.

The discussion in this chapter will center on assessment of environmental exposure as defined in the previous sentences. Consequently, there will be no discussion of *biological markers*, i.e., markers of internal dose, biologically effective dose, biological response, or susceptibility.[2] The use of such biological markers in epidemiology is increasing rapidly, and it is important to recognise that in several cases in which environmental exposures are complex and/or difficult to quantitate, the use of biological markers of internal dose offers clear advantages over trying to measure or estimate environmental exposure with sophisticated and costly methods (assessment of exposure to environmental lead is a case in point; measurement of lead in blood is a more effective and reliable way to assess integrated exposure to lead in many different media than measurement of lead in all of these media).

EXPOSURE ASSESSMENT IN ENVIRONMENTAL EPIDEMIOLOGY

Environmental epidemiology is the study of the distribution of human health and disease in relationship to exposure to environmental agents. Exposure assessment in environmental epidemiology therefore has a different and more limited aim than assessment of human exposure to environmental agents in general. Other aims of exposure assessment can be the development of effective control strategies, or control of compliance to environmental quality standards. In environmental epidemiology, exposure assessment must primarily fulfil the needs of the epidemiologist. The epidemiologist, who studies the effect of exposure to environmental agents on health and disease, needs measures of exposure that are valid and precise, so that the effect of exposure on disease can be estimated with minimal bias.

VALIDITY, PRECISION, AND BIAS

The *validity* of a measure of exposure refers to the agreement between this measure and the true exposure. If individual exposure to, for example, dust in the air is largely determined by whether or not there are smokers in the home, the validity of the dust concentration measured at one central monitoring site in an urban area as a measure of the true exposure to dust is poor. It is poor because it does not correlate well with personal exposure, and it is poor because it will *under*estimate dust exposure of subjects living with smokers. At the same time, it may *over*estimate dust exposure of subjects living in homes without smokers, due to the protection that being indoors offers to dust in ambient air. In addition, there may be differences in the composition of the dust that are biologically relevant. It is important to realise that a poor correlation between some measure of exposure and personal, true exposure is potentially a more serious problem than a systematic difference between measured and true exposure, because in the latter case the power of a study to detect a relationship between exposure and disease is not compromised. In quantitative terms, the detected relationship will still be biased; however, the magnitude and direction of the bias depend on the magnitude and the direction of the systematic difference between measured and true exposure.

The *precision* of a measure of exposure refers to its repeatability. Partly, repeatability is a technical issue related to characteristics of instrumentation and analytical procedures, but the errors of sampling and analysis are often much smaller than the true variability of exposure measures in time and space. Assessment of the precision of a measure of exposure calls for prior definition of the spatial and temporal characteristics of the true exposure variable. Often, exposure is being measured in more limited domains of time and space than the true exposure we are interested in. Some repetition of exposure measurements in time and space is needed to estimate the variability of interest. Errors of sampling and analysis are commonly expressed as coefficients of variation, i.e., the standard deviation of a series of measurements repeated in time and/or space divided by the series mean. This quantity is then usually multiplied by a factor of 100 to obtain the coefficient of variation. An important feature of the precision of exposure assessment for the epidemiologist is that the relative precision (variability of the individual exposure measure relative to the population variability) is more important than the absolute precision. This is related to the so-called exposure misclassification bias, which will be discussed in more detail after some general remarks about *bias* in epidemiology.

Bias in epidemiology refers to the difference between the estimated association between exposure and disease, and the true association. Three general types of bias can be identified: selection bias, information bias, and confounding bias.[3] For the purpose of this chapter, the discussion can be restricted to information bias, which occurs when the information about exposure and/or disease is wrong to the extent that the relationship between the two is no longer correctly being estimated. Errors in exposure assessment can lead to information bias. In addition to the notions of validity and precision of measures of exposure, the concepts of *differential misclassification* and *nondifferential misclassification* are often used in epidemiology to put these notions in the context of classification of study subjects into various categories of exposure. Differential misclassification means that the extent of misclassification is different for cases and controls. Nondifferential misclassification refers to a misclassification of exposure which is not different between cases and controls. The extent of nondifferential misclassification is related to both the precision of the exposure measurement and the magnitude of the differences in exposure within the population. If the differences in exposure are substantial, even a fairly imprecise exposure measurement would not lead to much misclassification of exposure. For this reason, it is important to judge the precision of exposure measurements always relative to the magnitude of the exposure variability within the population.

Nondifferential misclassification of exposure generally results in a *bias to the null* in estimates of relationships between exposure and disease.[4] Because nondifferential misclassification of exposure is directly related to the relative precision of the measurement of the exposure variable, a systematic investigation of the relative precision of the measurement of the exposure variable should ideally precede any study in environmental epidemiology. It seems, however, that this point has not received as much attention as it has in, for example, nutritional epidemiology.[5,6]

A practical illustration of the bias introduced by nondifferential misclassification of exposure is taken from Brunekreef et al.[7] In a study on the relationship between environmental exposure to lead and blood lead in children four different measures of lead on home floors were taken for each participating subject. An analysis of variance indicated that the variance component within homes was 1.33 times as large as the variance component between homes. Theory predicts that using one measurement of lead on home floors as an exposure variable would lead to an underestimation of the true regression coefficient of blood lead on lead exposure by a factor of $(1 + 1.33) = 2.33$.[8] Averaging the four separate measurements of lead exposure into one would lead one to expect an underestimation of $(1 + 1.33/4) = 1.33$.[9] Table 1 lists the results of five different regression analyses, in which the four separate measurements of lead exposure as well as their average were used as predictors of blood lead.

Table 1   Regression Coefficients of Blood Lead on House Dust Lead

Separate Measurements	Average
12.3	12.7
5.7	
9.0	
6.2	

Note: Coefficients in $\mu g/100$ ml per $\mu g/m^2$. Adapted from Brunekreef et al., *Am. J. Epidemiol.* 125, 892, 1987.

The results in the table indicate that the coefficients for each of the separate measurements were smaller than the coefficient obtained when the average exposure was used as predictor variable. From theory, we would expect a ratio of $2.33/1.33 = 1.75$ between the coefficients obtained with the average and the separate lead measurements, respectively. The observed ratio was: $12.7/\{(12.3 + 5.7 + 9.0 + 6.2)/4\} = 1.69$, close to the expected value.

After these more or less theoretical remarks, the remainder of this chapter will be devoted to some of the major problems in exposure assessment we face in environmental epidemiology today.

RETROSPECTIVE EXPOSURE ASSESSMENT
IN ENVIRONMENTAL EPIDEMIOLOGY

Often, the environmental epidemiologist is interested in the effect of environmental exposures on chronic diseases such as lung cancer. In view of the long periods of time that may pass before exposed subjects become diseased, case-control studies have been the study design of choice to investigate such exposures. Exposure then has to be assessed retrospectively, and there is a clear need for development and validation of retrospective exposure assessment methods. In occupational epidemiology, so-called job-exposure matrices have been developed to assist investigators in retrospectively assigning specific exposures to individual workers on the basis of descriptions of the jobs they held.[10] Development of methods to reconstruct environmental exposures retrospectively on the basis of residential histories is still in its infancy. Interesting examples can be found in recent studies on the development of lung cancer in relationship to exposure to radon[11] and air pollution.[12] Svensson and co-workers[11] conducted a case-control study on lung cancer in females in Sweden, comparing radon concentrations in previous dwellings between cases and controls. The study included 210 incident female lung cancer cases, 209 population controls, and 191 hospital controls. These 610 subjects had lived in a total of 3518 dwellings for more than 2 years. In a random sample of 303 of these, radon measurements were performed. These measurements were used to estimate lifelong cumulative radon exposures for all cases and controls. The results were analysed in conjunction with data on smoking, and the main results are summarised in Table 2. The data indicate that there was an increase in lung cancer risk with increasing exposure to radon, and that the increase was largely confined to the smokers in the population. This study is interesting in that it presents one of the few systematic attempts in the published literature to obtain quantitative, retrospective estimates of exposure to an environmental agent.

Table 2   Relative Risk Estimates for Smoking and Cumulative Radon Exposure

Smoking	<4500 Bq/m^3·years	4501–6000 Bq/m^3·years	>6000 Bq/m^3·years
Never	1.0	1.4	0.9
1–10 cig/d	2.3	4.8	6.5
>10 cig/d	6.8	12.3	15.9

Adapted from Svensson et al., *Cancer Res.*, 49, 1861, 1989.

Jedrychowski and co-workers[12] compared 1579 lung cancer deaths in Cracow, Poland, which were found in the death register over the period of 1980–1985 to a reference population of 1491 other deaths (excluding respiratory deaths), frequency matched for age and sex. Next of kin were interviewed in order to obtain information on individual smoking habits, occupation, and last residence. On the basis of existing records of air pollution concentrations (total suspended particulates [TSP] and sulfur dioxide [SO$_2$]) for the area, subjects' residences were classified into air pollution exposure categories. After adjustment for smoking and occupational exposures, the authors found that under conditions found in Cracow, air pollution may increase lung cancer risk, acting multiplicatively with smoking and occupational exposure. This is another example of a study in which the investigators tried to obtain historical data on environmental exposure as well as exposure to two different major confounders. This study was more limited than the Swedish radon study in the sense that the investigators did not attempt to obtain information about previous residences.

MICRO-ENVIRONMENTAL EXPOSURES
IN ENVIRONMENTAL EPIDEMIOLOGY

A major advance in environmental (especially air pollution) epidemiology of the past 15 years has been fueled by the realisation that human exposure to pollutants in micro-environments may be very different from exposure in the general environment. In moderate or cold climates, human beings spend the vast majority of their time indoors, and cumulative exposure to a range of substances is clearly determined by indoor rather than outdoor exposures. As the microenvironments where people spend time are different for each subject in a population, exposure assessment that tries to take these peculiarities into account is clearly a complicated enterprise. There have been two major approaches to solving this problem, one being the development of instrumentation suitable for microenvironmental and personal monitoring and the other being the development of sophisticated exposure models. Several investigators have employed miniaturised instruments to measure components such as nitrogen dioxide (NO$_2$) and respirable suspended particulate matter in indoor air, in successful attempts to obtain individual exposure assessments in large-scale epidemiologic studies.[13,14] Others have employed modelling approaches, using data on indoor and outdoor sources, relationships between sources, source use and micro-environmental monitoring data, and human time-activity patterns to model personal exposure to (mostly air) pollutants.[15] Clearly, information coming from many different sources is needed to successfully model human exposure, and standardisation of methods to collect this information is important to improve comparison of study results. A significant attempt to achieve such standardisation for making inventories of indoor air pollution sources was published by Lebowitz et al.[16] A recent monograph published by the U.S. National Academy of Sciences gives a detailed overview of the developments in the field of human exposure assessment to airborne pollutants.[17] A discussion of exposure assessment specifically in the framework of air pollution epidemiology is given in a recent report produced by a European expert group.[18] It is important to consider that there is a trade-off between measuring and modelling personal exposure in environmental epidemiology studies; measuring exposure is invariably more expensive than modelling exposure; thus there is always the possibility that the gains in validity and precision associated with measuring exposure are offset by the losses in the power of a study to detect significant relationships between exposure and effect.[19]

MULTI-MEDIA EXPOSURES IN ENVIRONMENTAL EPIDEMIOLOGY

Many pollutants are so dispersed in the environment that they can reach the human body through a complex variety of environmental pathways. Pollutants such as cadmium, lead, dioxin, and PCBs are just a few of the many examples. Measuring and modelling of integrated exposure to such substances are difficult at best, and measurement of biological markers of dose will be the preferred strategy in many cases for those pollutants for which such markers are available. Measurement of biological markers of dose (such as the concentration of lead in blood) does not usually tell us which environmental sources and pathways are dominating exposure, however. The development of multimedia measurement or modelling approaches is necessary to resolve this question.[20] An example of a detailed human exposure assessment to dioxin is given by Travis and Hattemer-Frey.[21]

Sometimes, specific tracers can be used to trace environmental pathways from pollutant sources to human beings. The stable isotope ratio of lead has been used to identify lead sources of exposure of children living in California in one study.[22] An ingenious example of using the isotope ratio of lead to trace environmental exposure was the Turin Isotopic Lead Experiment.[23] For 4 years (1975–1979), the Pb-206/Pb-207 isotopic ratio of lead in gasoline was changed from 1.18 to 1.06 in a large region surrounding and including Turin, Italy, in order to document the importance of gasoline lead for the total human exposure to lead. For a number of years, the isotopic ratio was measured in various environmental media and in human blood, and a number of important conclusions could be drawn from this experimental study in environmental epidemiology with regards to the origins of lead in the human body.

Another example of studies trying to elucidate complex environmental pathways is found in some recent studies that have tried to document soil ingestion by young children by measuring typical soil constituents (such as titanium, silicon, and aluminum) in soil and in human excreta.[24,25] Although these constituents can be found in the human diet due to other sources as well, their quantitative combination in soil is fairly stable and unique, so that reasonable estimates of the upper bound of soil ingestion by children could be made.

CONCLUDING REMARKS

Exposure assessment in environmental epidemiology is complicated due to the many potential pathways and to the many microenvironments in which people spend their time. In analytical studies in environmental epidemiology, measures of exposure are needed that are optimal with respect to the power of a study to detect a significant relationship between exposure and disease, and with respect to the validity of this relationship. When designing studies in environmental epidemiology, the investigator ideally would like to have answers to the following questions:

1. How well does a measure of exposure correlate with true exposure?
2. How large is the ratio of the variance *within* subjects, and the variance *between* subjects of a measure of exposure?
3. Taking into account the unit cost associated with obtaining a measure of exposure, which of a number of competing measures should be chosen in order to maximise the power of a study to detect a significant relationship between exposure and disease?

Exposure assessment in the environmental epidemiology of the future will undoubtedly benefit from specific studies targeted at answering these questions.

REFERENCES

1. Sexton, K. and Ryan, P. B., Assessment of human exposure to air pollution: methods, measurements and models, in *Air Pollution, the Automobile, and Public Health,* Watson, A. Y., Bates, R. R., and Kennedy, D., Eds., The Health Effects Institute, National Academy Press, Washington, D.C., 1988, 207-38.
2. Hulka, B. S., Wilcosky, T. C., and Griffith, J. D., *Biological Markers in Epidemiology,* Oxford University Press, New York/Oxford, 1990.
3. Rothman K. J., *Modern Epidemiology,* Little, Brown and Company, Boston/Toronto, 1986.
4. Kelsey, J. L., Thompson, W. D., and Evans, A. S., *Methods in Observational Epidemiology,* Oxford University Press, New York/Oxford, 1986.
5. Chalmers, F. W., Clayton, M. M., Lorraine, O., et al., The dietary record — how many and which days? *J. Am. Diet. Assoc.,* 28, 711, 1952.
6. Willett W., *Nutritional Epidemiology,* Oxford University Press, New York/Oxford, 1990.
7. Brunekreef, B., Noy, D., and Clausing, P., Variability of exposure measurements in environmental epidemiology, *Am. J. Epidemiol.,* 125, 892, 1987.
8. Snedecor, G. W., and Cochran, W. G., *Statistical Methods (6th edition),* The Iowa State University Press, Ames, Iowa, 1967.
9. Liu, K., Stamler, J., Dyer, A. et al., Statistical methods to assess and minimize the role of intra-indivdual variability in obscuring the relationship between dietary lipds and serum cholesterol, *J. Chron. Dis.,* 31, 399, 1978.

10. Pannett, B., Coggon, D., and Acheson, R. E. D., A job-exposure matrix for use in population based studies in England and Wales, *Br. J. Ind. Med.*, 42, 777, 1985.

11. Svensson, C., Pershagen, G., and Klominek, J., Lung cancer in women and type of dwelling in relation to radon exposure, *Cancer Res.* 49, 1861, 1989.

12. Jedrychowski, W., Becher, H., Wahrendorf, J., and Basa-Cierpialek, Z., A case-control study of lung cancer with special reference to the effect of air pollution in Poland, *J. Epidemiol. Commun. Health,* 44, 114, 1990.

13. Quackenboss, J. J., Krzyzanowski, M., and Lebowitz, M. D., Exposure assessment approaches to evaluate respiratory health effects of particulate matter and nitrogen dioxide, *J. Exp. Anal. Environ. Epidemiol.*, 1, 83, 1991.

14. Neas, L. M., Dockery, D. W., Ware, J. H. et al., Association of indoor nitrogen dioxide with respiratory symptoms and pulmonary function, *Am. J. Epidemiol.*, 134, 204, 1991.

15. Duan, N., Stochastic micro-environmental models for air pollution exposure, *J. Exp. Anal. Environ. Epidemiol.*, 1, 235, 1991.

16. Lebowitz, M. D., Quackenboss, J. J., Kollander, M., Soczak, M. L., and Colome, S.D., The new Standard Environmental Inventory Questionnaire for Estimation of Indoor Concentrations, *J. Air Waste Manage. Assoc. JAPCA*, 39, 1411, 1989.

17. National Research Council, *Human Exposure Assessment for Airborne Pollutants — Advances and Opportunities,* National Academy Press, Washington, D.C., 1991.

18. Commission of the European Communities, Exposure Assessment — COST 613/2 Report Series on Air Pollution Epidemiology, Report no. 1, Brussels, 1992.

19. Noy, D., Brunekreef, B., Houthuijs, D., Boleij, J. S. M., and Koning, R. de., The assessment of personal exposure to nitrogen dioxide in epidemiologic studies, *Atmos. Environ.*, 24A, 2903, 1990.

20. Stevens, J. B. and Swackhamer, D. L., Environmental pollution: a multimedia approach to modelling human exposure, *Environ. Sci. Technol.*, 23, 1180, 1989.

21. Travis, C. T. and Hattemer-Frey, H. A., Human exposure to dioxin, *Sci. Total Environ.*, 104, 97, 1991.

22. Yaffe, Y., Flessel, C. P., Wesolowski, J. J., et al., Identification of lead sources in California children using a stable isotope ratio technique, *Arch. Environ. Health,* 38, 237, 1983.

23. Fachetti, S. and Geiss, F., Isotopic Lead Experiment — Status Report, Commission of the European Communities Report EUR 8352 EN, 1982.

24. Calabrese, E. J., Barnes, R., Stanek, III E. J., et al., How much soil do young children ingest: an epidemiologic study, *Reg. Toxicol. Pharmacol.*, 10, 123, 1989.

25. Wijnen, J. H. van, Clausing, P., and Brunekreef, B., Estimated soil ingestion by children, *Environ. Res.*, 51, 157, 1991.

15 | NEEDS AND REQUIREMENTS FOR UNDERGRADUATE AND GRADUATE TRAINING IN ENVIRONMENTAL AND OCCUPATIONAL EPIDEMIOLOGY

Wieslaw Jedrychowski

CONTENTS

INTRODUCTION

Training in modern medicine aims at teaching in detail the structure and function of the human organism, both in health and in disease. On the other hand, it also aims at revealing the factors controlling normal or abnormal human development both of a single human being and of the population. The first field of studies involves mainly molecular biology, biochemistry, biophysics, genetics, and immunology, and also other branches of human physiology and pathology, which are the basis for modern diagnostic and therapeutic medicine. The second is based mainly on epidemiology, which deals with factors determining the frequency and distribution of diseases and other health phenomena in human populations.

TRAINING OF UNDERGRADUATE MEDICAL STUDENTS

The aims of teaching environmental and occupational epidemiology to undergraduate medical students are numerous. An important contribution that epidemiology makes is the recognition that the health of the community is not determined only by the traditional clinical specialists of medicine. Each student of medicine must understand the setting in which disease and health care behaviour occurs; and those demographic, occupational, social, and community aspects that define how people get health care and how they get diseases. He should be able to identify the risks to which patients are subject, describe the characteristics that define such risks, and have at least an understanding of the strategy of effectiveness of medical therapy. It is desirable that every physician, before he goes into practice, should participate in a clinical trial just to know how meaningful it is.

In other words, environmental and occupational epidemiology should give medical students an understanding of the interactions between man and environment, between the causes of disease, of the frequency of diseases in populations, and of how and why these frequencies are changing. It should teach a critical attitude to both the effectiveness and efficiency of medical care. As it gives medical students a broader horizon and greater perspective, it is important to put greater emphasis on the educational role of epidemiology in the medical curriculum. As nowadays modern society turns from an almost exclusive interest in the care of the individuals who are ill toward an organised effort to preserve the health of a population, it is imperative that all physicians have a good knowledge of epidemiology. The practitioner should know not only about diagnosis and treatment of a given disease, but also something about evaluation of medical care in terms of processes or outcomes. In fact, perhaps instead of a course in epidemiology there should be a course in health care into which the epidemiological methods and skills are presented in such a way that he will understand how they are used, and why he needs these skills to be an effective practitioner. Consideration of the reasons for teaching epidemiology indicates that every medical student should be exposed to it and that it should not merely be an elective subject among many others.[1] The undergraduate medical curriculum suffers from a very crowded program, aggravated by constant demands from increasing allocations of teaching hours by teachers of the various disciplines. However, their epidemiological training should be introduced as early as possible in the curriculum and possibly divided into two parts, theoretical and practical. First, there should be a short, systematic course on the principles of descriptive, analytical, and experimental epidemiology; and this should be followed by practical exercises or small field epidemiological projects related to environmental problems. The exact duration of the course would depend on the particular curriculum.

There is a need to give considerable thought to exactly what amount of epidemiology is needed by any medical practitioner who will be engaged in the practice of medicine and to consider, from the point of economy and efficiency, ways to teach epidemiology to medical students in an attractive manner. In some countries there has been a definite shift among medical students away from interest in the community. This is perhaps a reflection of the way they are chosen. It may be also related to the fact that they want to practice medicine along the traditional lines of the one-to-one doctor-patient relationship of practitioners of the past and this serves as their present-day model. If this is so, we have to consider how these attitudes may be changed in the desired direction during the undergraduate epidemiology courses.

According to teaching guidelines established by Lowe and Kostrzewski,[2] in more specific terms undergraduate training in epidemiology should ensure that every doctor should be able to:

1. Describe the population structure of the country in which he will be working and appreciate the relation this has to the health problems of that country
2. Calculate simple rates, such as birth and death rates, or specific morbidity and mortality rates; perform simple analyses of morbidity and mortality data, according to person, place, and time
3. Collect and analyse hospital and community health data and present them clearly in tables and graphs
4. Complete notification and registration forms and other medical records required for statutory purposes in the proper way, using the International Classification of Diseases, or in special instances, other standardised classifications
5. Give an account of the epidemiology of the common diseases and apply this knowledge to the daily management of his patients
6. Understand how hypotheses are formulated and tested by epidemiologists and how epidemiological knowledge is applied to the control of communicable and noncommunicable diseases and to the evaluation of therapeutic and preventive procedures
7. Understand some of the problems associated with the design and interpretation of epidemiological investigations
8. Understand some of the problems associated with the design and interpretation of epidemiological investigations
9. Read the epidemiological literature critically and identify possible sources of bias and error in published data
10. Understand some of the more important uses of the computer in epidemiology

GRADUATE TRAINING IN ENVIRONMENTAL AND OCCUPATIONAL EPIDEMIOLOGY

With the increasing awareness that the maintenance of health in populations is an intricate ecological problem, there is also a growing need for professional epidemiologists who would comprehend the great importance of preventive medicine, understand the nature of environmental and occupational hazards, and have the knowledge how to organise the delivery of medical care in an efficient way. Therefore, professional epidemiologists have to assist in recognising the need for epidemiological information in the development of health risk assessment and any arrangement of health services seeking to balance equity of access to care, fair shares in the distribution of resources, and responsible moderation of costs.[3,4]

The need for epidemiologists, in the future, is likely to continue to increase with the increasing prominence of environmental and occupational epidemiology. It is due to the increasing contamination of the environment that causes the development of environment-related health problems. As many countries now are more often devoting governmental and private resources to the control of these hazards, effective control programs require more well-trained staff at different levels.

As health problems of the populations and their determinants change over time, epidemiology must keep pace with these changes and developments. Hence, there is a strong need to prepare a new generation of epidemiologists. Together with this challenge there is a great need to train clinicians, administrators, and policymakers with the epidemiological viewpoint and philosophy. It is obvious that the objectives of graduate teaching in epidemiology are different from those of undergraduate teaching and a knowledge of epidemiology is desirable for a wide range of health workers. These workers include those cited below.

Professional epidemiologists devote all or the greater part of their time to the use of epidemiology in practice, teaching, and research. This may be in communicable diseases, noncommunicable diseases, or evaluation of the effectiveness and efficiency of health services. They are expected to design and conduct epidemiological studies and be familiar with the uses of all important statistical methods.

Health services administrators and other public health workers should have a thorough ground in the uses of epidemiological methods for tackling health problems in the community, including the evaluation of health services. They should be able to collect and interpret data and assist in the design of epidemiological studies.

Clinicians and graduates from other specialities should be able to recognise the role of epidemiology and biostatistics as they relate to their own work. They should also be able to read the epidemiological literature critically. Clinicians who seek epidemiological instruction fall into two distinct groups. Members of the first group want to learn about the origins and control of disease in order to care for their patients more efficiently in their day-to-day practice. Members of the other group have the broader interest of learning how to plan their own and their colleagues' work in relation to the overall structure of health services and health needs in their area. By acquiring some insight into epidemiology they become more competent in committee work and planning.

Paramedical and auxiliary personnel are expected to assist the professional epidemiologist in his activities and to give technical assistance to other graduates who are using epidemiological methods. They should be acquainted with the basic principles of the subject and with the practical use of data handling methods and statistical calculations.

Professional epidemiologists, other public health workers, clinicians, and other professional graduates may be trained in any suitable medical school, school of public health, postgraduate medical institute, or research institute. Successful short courses may also be run in non-teaching hospitals and community health centres. In general, epidemiological training should be given in cultural circumstances similar to those in which the trainee will eventually work; although for advanced instruction, it may be necessary to attend specialist institutes. The International Epidemiological Association and WHO have also been running a series of training courses and special scientific meetings which have contributed to epidemiological education of medical graduates. Some of these have taken the form of intensive seminars lasting a few days and have been held in the host countries.

CURRENTLY AVAILABLE COURSES OF EPIDEMIOLOGY IN EUROPE

The analysis of the current teaching courses and seminars of epidemiology in Europe has been based on the Inventory of Environmental and Occupational Epidemiology Courses[5] compiled by WHO Environmental Epidemiology Division of Environmental Health (GEENET, WHO, Geneva, 1992). It refers to courses planned over the period 1992-1993 (Table 1). Out of 43 postgraduate

Table 1   Number of Courses in Each Country from Europe

Belgium	2	Czechoslovakia	2
Denmark	1	Germany	2
France	3	Greece	1
Ireland	3	Italy	8
Poland	1	Portugal	1
Russia	1	Spain	1
Sweden	4	Switzerland	1
The Netherlands	1	Turkey	1
United Kingdom	10		

courses to be held in Europe over the given period, 39 units are to be held in Western Europe, and only four in central or eastern European countries. The courses are organised mostly on a regular basis (77%). In the eastern part of Europe there are only courses of short duration (one month or less), while in the western European countries about half of them is of short duration, and 13% lasts longer than one year. Neither of the courses in the eastern European countries offers diploma nor degree. In Western Europe 23% of the courses leads to a diploma and 33% leads to other degrees. On average, half of the courses in Europe is based on the full-time commitment of students. Most of the courses are at post-basic level (61%) and 36% of them is of higher professional level. Infectious diseases are covered in 90% of the courses, while chronic or non-infectious diseases are covered in 60% of the courses. Statistical methods and survey design are presented in 93% of the courses, methods of field investigation in 74%, risk assessment and risk management in 79%, and formulation of policy in 60%.

In addition to that, GEENET[6] also collected information on local perception of environmental problems, and perceived needs for training in environmental epidemiology in European countries and in other parts of the world. As might be expected, the environmental problems considered as very serious differed depending upon country. In the European region traffic accidents were the most frequently reported very serious problem; and this was followed by air pollution from cars, air pollution from power plants and industry, toxic waste disposal sites, and pollution from toxic waste disposal or spills (Table 2).

Table 2   Ways in Which Training Could Be Improved (By Areas)

Improvement Needed	University Tech. Schools (%)	Short Courses Seminars %
Practical or field training	18.7	5.5
Environmental health	8.1	1.8
Epidemiology	6.0	2.3
Research	3.2	0.2
Prevention	2.3	0.2
Expand training time	39.0	18.0
Develop new degree/diploma	10.2	0.0
Integrate curricula	12.9	1.6
Establish school of public health	3.2	0.0
Medical student training	9.7	0.7
Better use of training material	8.8	5.1
Improve facilities	4.8	1.2
Increase financial sources	5.5	3.5
Improve quality of faculty	6.9	2.5
Improve quality of curricula	5.3	2.3
Increase visibility of public health as a field of study	4.2	1.4
Continuing education for professionals	1.6	14.1
Training for inspectors	1.2	2.5
Training for workers	0.2	6.5
Public education	0.2	2.3
Intercountry cooperation	1.6	3.7
Occupational health	2.8	0.9
Other	25.6	19.4

Based upon the data published in Assessment of Training Needs in Environmental and Occupational Health. GEENET, WHO, Geneva, 1992.

For both university/technical schools and short course/seminars, expanded training time was the most frequently reported need expressed (Table 3). This included increasing teaching hours, adding to the number of courses, and increasing the number of teachers. It was postulated to put a more practical emphasis on field training, integrating curricula from different departments or disciplines, developing new degree or diploma programs, and new curriculum in medical student training. The fields most frequently reported as needing additional training (Table 4) were occupational health, toxicology, and epidemiology. The respondents stressed the need for WHO support in developing the training programs in environmental epidemiology. They wanted aid in obtaining training texts or other teaching materials, and in facilitating international co-operation in training and teaching epidemiology to undergraduate and postgraduate students. In each medical school the number of teachers in epidemiology is usually small. This makes contacts with other medical schools highly desirable.

Table 3   Percent Reporting Very Serious Environmental and Occupational Problems in European WHO Region

Condition	%
Worker exposures to pesticides	8.3
Worker exposures to other chemicals	11.0
Air pollution from power plants and industry	19.3
Urban air pollution from motor cars	16.5
Pollution from agricultural chemicals and pesticides	11.9
Pollution from toxic waste disposal or spills	11.0
Traffic accidents	32.1
Inadequate sanitation and sewage disposal	13.8
Inadequate garbage disposal	9.2
Food contaminated by pesticides and/or chemicals	1.8
Food contaminated by pathogens	4.6
Household pesticides and/or other chemicals	1.8
Contamination of drinking water by chemicals/pesticides	6.4
Contamination of drinking water by pathogens	3.7
Local water pollution from industry	8.3
Toxic waste disposal site	12.8
Indoor air pollution from burning coal, petroleum, etc.	0.9
Floods, earthquakes, fires, and other disasters	2.8
Unhealthy and/or unsafe housing	3.7
Childhood lead exposure	3.7

Based upon the data published in Assessment of Training Needs in Environmental and Occupational Health. GEENET, WHO, Geneva, 1992.

The other problems mentioned by respondents were concerned with providing training for local teachers, and development of training curriculum guidelines (Table 5). The general feeling of the respondents was that the training in environmental epidemiology and occupational health should be improved in such a way that it meets the challenge of the current needs.

Problems and issues for consideration among those who are responsible for teaching programs in universities and technical schools are as follows:

**Table 4   Percent Reporting Fields in Which Additional
Training Is Necessary in European WHO Region**

Field	%
Epidemiology (including biostatistics)	19.3
Risk assessment	3.7
Exposure assessment/monitoring	3.7
Toxicology (including pesticides)	11.0
Occupational safety	0.0
Environmental and occupational injuries	5.5
Environmental and occupational diseases	2.8
Environmental health not otherwise specified	7.3
Occupational health not otherwise specified	11.0
Toxic waste management	1.8
Solid waste/sewage management	0.9
Food contamination (chemicals or pathogens)	2.8
Health education and community services	3.7
Water pollution	4.6
Air pollution	7.3
Soil contamination and other pollution unspecified	4.6
Other	25.7

Based upon the data published in Assessment of Training
Needs in Environmental and Occupational Health. GEENET,
WHO, Geneva, 1992.

**Table 5   How Could WHO Best Support Efforts to Improve
Training in Countries of European WHO Region**

Field	%
Develop training texts/materials	65.1
Develop training curriculum guidelines	45.0
Provide training for local teachers	34.9
Facilitate cooperation in training between countries	56.0

Based upon the data published in Assessment of Training Needs
in Environmental and Occupational Health. WHO, GEENET,
Geneva, 1992.

1. How can contemporary problems of medicine, health, and disease be understood better through education in environmental epidemiology?
2. At what point in the continuum of education of professional medical practice
   is environmental epidemiology most effectively taught and in what settings?
3. What is the desirable amount of knowledge and what kind of skills are the
   medical students to acquire in the course of epidemiology education?
4. What kinds of professionals can most effectively teach the principles,
   methods, and applications of environmental epidemiology?
5. How should we deal with the challenge to prepare a new generation of
   professional epidemiologists able to cope with contemporary health hazards, preventive measures, and delivery of medical care services?
6. Can we set up international guidelines for courses in training of undergraduate and postgraduate students as well as professionals in the field of
   environmental epidemiology?
7. How should local teachers of epidemiology be trained and how should
   international collaboration resources materials for teaching of environmental epidemiology be developed?

REFERENCES

1. White, K. L. and Henderson, M. M. (Ed.)., *Epidemiology as a Fundamental Science. Its Uses in Health Services Planning, Administration, and Evaluation.* Oxford University Press, Oxford, 1976.
2. Lowe, C. R. and Kostrzewski, J., *Epidemiology: A Guide for Teaching Methods.* Churchill Livingstone, IEA. 1973.
3. Abel-Smith, B., *Value for Money in Health Services.* Heinemann, London, 1976.
4. Holland, W. W. and Gilderdale, S. *Epidemiology and Health.* Henry Kimpton Publishers, London, 1977.
5. Inventory of Environmental and Occupational Epidemiology Courses. WHO Environmental Epidemiology Division of Environmental Health. GEENET, WHO, Geneva, 1992.
6. Assessment of Training Needs in Environmental and Occupational Health. GEENET, WHO, Geneva, 1992.
7. Vanderschmidt, H. F., Koch-Weser, D., and Woodbury, P. A. (Ed.). *Handbook of Clinical Prevention.* Williams & Wilkins, London-Sydney, 1987.

INDEX

INDEX

A

Acetylator phenotype, 205
Acid aerosols
  air pollution, 6–7, 12
  carcinogens, 12
Actinolite, See Asbestos
Acute accidents, 190
Acute lymphoblastic leukemia (ALL)
  epidemiological evidence, 136
  radon, 25
Acute myeloid leukemia (AML)
  electromagnetic fields, 103–104
  radionuclides, 66
  radon, 25, 27
Adenocarcinoma, 44
Adult onset respiratory hypersensitivity, 51
Aerosols, 6–7
Affected population size, 58
Aflatoxins, 131
Agency for Toxic Substances and Disease
    Registries (ATSDR), 49–52
Aggregate statistics, 59
Agricultural chemicals
  drinking water contaminants, 68–69, 74
  multiple myeloma, 136
  nervous system neoplasms, 135
  non-Hodgkin's lymphoma, 136
Air pollution
  acid aerosols, 6–7
  carbon monoxide, 8–9
  carcinogens, 10–11
    acid aerosols, 12
    asbestos, 11
    benzene, 11
    diesel exhaust, 12
  hazardous waste, 54–57
  introduction, 3–4
  lead, 9–10
  lung cancer, 148–51, 156
  nitrogen oxides, 7–8
  ozone, 4–6
  particulates, 6–7
  priority research needs, 12–14
  stomach cancer, 131
  sulfur dioxide, 6–7
Alcohol drinking
  breast cancer, 133
  esophageal cancer, 130
  kidney cancer, 135
  laryngeal cancer, 132
  liver cancer, 131
  oral cavity cancer, 130
  pancreatic cancer, 132
Aldehydes, 69
Aldrin, 85
Alkali, 136
Ambient pollution concentration records, 54
4-Aminobiphenyl-hemoglobin adducts, 205
Amitrole, 85
Amosite, See Asbestos
Anemia
  benzene, 11
  lead, 10
Aneuploidy, 45
Anginal pain, 8
Angiosarcoma, 56, 131
Anthophyllite, See Asbestos
Aplastic anemia, 11
Arsenic
  drinking water contaminants, 64, 66, 73,
    75
  lung cancer, 149
  pesticides, 85–86
Arteriosclerosis, 9
Artificial sweeteners, 134
Aryl hydrocarbon hydroxylase, 156
Asbestos
  air pollution, 11
  carcinogenicity evidence
    epidemiological evidence, 40
    experimental systems, 39